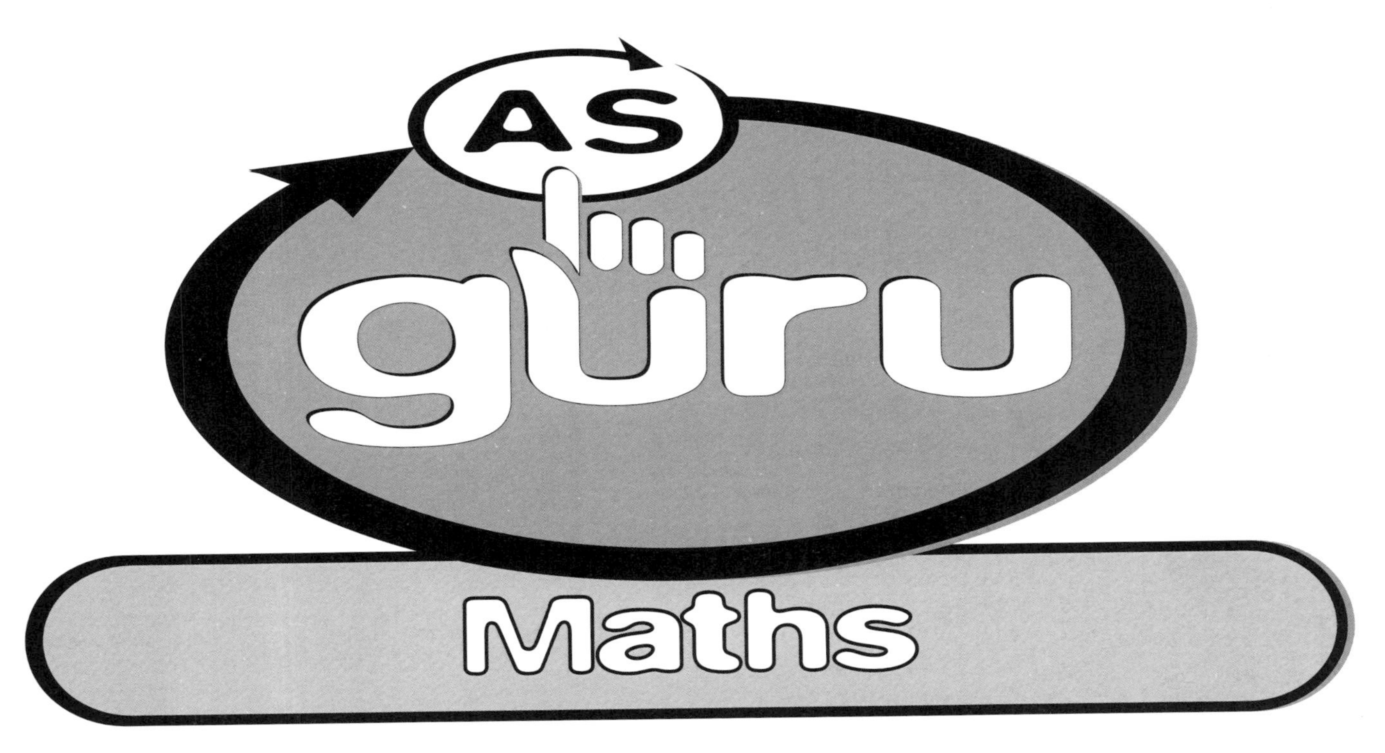

Stephen Hodges
With expert input and advice by Jean Matthews

With thanks to Charles Parker and Yvonne Gostling for additional questions

Published by BBC Educational Publishing, BBC White City, 201 Wood Lane, London W12 7TS

First published 2001

Designed by Steve Hollingshead

Illustrations by Hardlines

Printed in Great Britain by sterling press, Northamptonshire

ISBN 0 563 54234 9

To place an order, please telephone Customer Services on 01937 541001 (Monday – Friday, 0800 – 1800) or write to BBC Educational Publishing, PO Box 234, Wetherby, West Yorkshire, LS23 7EU.

Visit the BBC Education website at: www.bbc.co.uk/education

Contents

Introduction

From GCSE to AS Level maths is a big leap. You may feel as if you have entered a different world. This book aims to be your help and guide.

About this book

In mathematics – unlike in most other subjects – there are very few facts to learn. Instead, you learn techniques, which can then be applied to a wide range of problems. Learning mathematics must be an active process. A good deal of practice in solving problems is essential.

This book is divided into three main sections: Pure maths, Mechanics and Statistics. Each section goes through the main techniques you will need to know and shows worked examples to help consolidate the theory. The solutions are broken down into simple steps to make them easier to follow. Try to work through these examples for yourself. Find a comfortable desk and chair, equip yourself with a pencil and pad, and let the book guide you through the mathematics.

Getting started

The greatest difficulty in answering a question is often getting started. Which technique should you use? The more questions you try, the more clearly you can see which techniques would be most appropriate and the easier it becomes. Success in mathematics is more about perspiration than inspiration!

Practice

Having learnt a technique, the next thing to do is practise it. At the end of each of the three main sections of this book is a comprehensive set of practice questions. They have been specially written by experienced teachers and examiners to give you plenty of opportunity to use the techniques you have learnt again and again. Look back at the relevant place in the section you are working on if you have difficulty with any of the questions. All the answers are included at the back of the book.

GURU TIP
Tip boxes like this appear throughout the book to give you good ideas and hints about maths.

WEB TIP
Where there's a close link with the AS Guru™ website you'll see a box like this.

Information boxes like this give you extra facts and details you should know.

Features in this book

- Throughout the book, Guru Tips appear in the margin. These are tips and reminders from the experts which will help you to learn, practise and revise your AS Level maths.
- Specific links to the website are pointed out in the margins, too.
- There are also markers to show where you could possibly use some of your AS level work to help compile your Key Skills portfolio. There's more about this on page 6.
- The most vital facts and information appear in green boxes within the text, so you can pick them out instantly and refer to them quickly when you're skimming the book to find out something you need to know, or to revise.
- Answers are also picked out in green, so you can find them easily. You should always try to indicate your answers so they stand out clearly from your working. Double underlining is a good method.

Using the AS Guru book

You can either work through the book from the start, to make sure you have covered all the ground, or just work on specific sections as you cover that topic in class and want to reinforce what you've learnt.

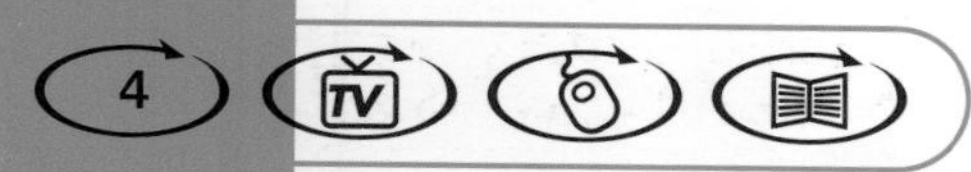

The book also makes a good revision guide – try working through it when it comes to exam time to make sure you get plenty of practice. This book cannot cover every single possible question that might arise, however. The work you do in class, homework and extra, independent studies of your own are all important to getting a good grade in your AS Level maths exams.

The full AS Guru™ service

AS Guru™ is a cross-media guide to AS Level Maths (and lots of other subjects) across books, television programmes (there are no television programmes for maths, however) and an extensive website. The media are complementary and, if you can, you should go to the website for more worked examples and explanations of topics in this book.

Visit the AS Guru™ website at http://www.bbc.co.uk/education/asguru/maths

Other AS Guru™ subjects available now across books, websites and television are:

GURU TIP
There's also an AS Guru™ Study Skills video available for £9.99, Code: EDUC 7126

BIOLOGY

Book: ISBN 0 563 54241 1
website: http://www.bbc.co.uk/education/asguru/biology

Television:	Programme	Transmission times
1:	Biological molecules, Organs and systems, and Ecology	Sat/Sun 10/11th Feb, 2001 0300-0500 Sat/Sun 29/30th April, 2001 0300-0500
2:	Cell biology and Genetics	Fri/Sat 16/17th Feb, 2001 0300-0500 Sun/Mon 29/30th April, 2001 0200-0400

WEB TIP
On the website, you'll find interactive exercises to help you understand – and which are fun to do!

There's also a section on Study Skills which will help with some background to AS Level maths, particularly algebra.

ENGLISH

Book: ISBN 0 563 54235 7
website: http://www.bbc.co.uk/education/asguru/english

Television:	Programme	Transmission times
1:	What is text? Genre, Context, Shakespeare	Sat/Sun 3/4th Feb, 2001 0300-0500 Sun/Mon 22nd/23rd April, 2001 0200-0400
2:	Different interpretations, integrating language and literature, speech and writing, systematic language frameworks, prose set text, context, original writing	Fri/Sat 9/10th Feb, 2001 0300-0500 Fri/Sat 27/28th April, 2001 0300-0500

GENERAL STUDIES

Book: ISBN 0 563 54236 5
website: http://www.bbc.co.uk/education/asguru/generalstudies

Television:	Programme	Transmission times
1:	Culture	Sat/Sun 27/28th Jan, 2001 0400-0500 Fri/Sat 20th/21st April, 2001 0400-0500
2:	Science and Economics	Fri/Sat 2nd/3rd Feb, 2001 0300-0500 Sat/Sun 21st/22nd April, 2001 0300-0500

GURU TIP
Remember to double-check the dates of your module exams, and that you've got a spare battery for your calculator.

Awarding bodies

There are some variations in topics between examination awarding bodies. Check with your teacher if you're not sure whether any topic in AS Guru will appear in your exam. You may have an option to submit coursework as part of some modules, too.

Key skills

Your work at AS Level can help you to gain a new, extra Key Skills qualification. This award shows that you've developed important life skills that you'll need after your studies; skills that are in demand by employers and Further Education establishments. There are six Key Skills, but only three of them – Communication, Information Technology and Application of number – contribute to the Key Skills qualification.

To gain a Key Skills qualification, you need to collect a portfolio of evidence, which demonstrates your level of competence in the three main Key Skills areas. There are three levels of award (1-3) and you should aim to meet the criteria laid down for level 3.

The chart below shows you the skills you must demonstrate in the three main Key Skills areas, and what's required to fulfil each one. Where there are opportunities to demonstrate these skills through your work in AS Level maths, these are shown in this book.

C3	**Communication level 3**
C3.1a	Contribute to a group discussion about a complex subject.
C3.1b	Make a presentation about a complex subject, using at least one image to illustrate complex points.
C3.2	Read and synthesise information from two extended documents that deal with a complex subject. One of these documents should include at least one image.
C3.3	Write two different types of documents about complex subjects. One piece of writing should be an extended document and include at least one image.
N3	**Application of number level 3**
N3.1	Plan, and interpret information from two types of sources including a large data set.
N3.2	Carry out multi-stage calculations to do with: amounts and sizes; scales and proportion; handling statistics and rearranging and using formulae. Work with a large data set on at least one occasion.
N3.3	Interpret results of your calculations, present your findings and justify your methods. You must use at least one graph, one chart and one diagram.
IT3	**IT level 3**
IT3.1	Plan and use different sources to search for, and select, information required for two different purposes.
IT3.2	Explore, develop, and exchange information and derive new information to meet two different purposes.
IT3.3	Present information from different sources for two different purposes and audiences. Your work must include at least one example of text, one example of an image and one example of numbers.

Introduction – mathematical proof

What is pure mathematics?

The mathematics you learnt at GCSE was largely concerned with measuring, counting and plane geometry, with some algebra.

In pure mathematics you will deepen your knowledge of number, functions, algebra and proof, from which you will develop techniques like differentiation and integration (calculus).

Essentially, calculus is the study of motion and change. We live in a changing world. The techniques of calculus may be applied to model these changes – for example, the spread of an epidemic, or the motion of a ball through the air.

In statistics or mechanics, you apply the techniques rigorously developed in pure to model real-life situations.

Structuring a mathematical argument

Precision is very important in mathematics. Arguments should be logically structured. The language of mathematical arguments is algebra. You need to know the following symbols:

A $\Rightarrow$ B: Statement A implies statement B. ('If A then B.') A is a **sufficient** condition for B.

A $\Leftarrow$ B: Statement A is implied by statement B. B is a **necessary** condition for A.

A $\Leftrightarrow$ B: The implication works both ways.

The statement A $\Leftarrow$ B (or B $\Rightarrow$ A) is called the **converse** of the statement A $\Rightarrow$ B.

EXAMPLE Which of the following statements are true?

(a) An integer n is divisible by 5 $\Rightarrow$ the number ends in 0.

(b) An integer n is divisible by 5 $\Leftarrow$ the number ends in 0.

(c) The angles of a triangle are equal $\Leftarrow$ the sides of a triangle are equal.

(d) The angles of a triangle equal $\Rightarrow$ the sides of a triangle are equal.

SOLUTION (a) The fact that a number is divisible by 5 does not imply that it ends in 0. 25 is a **counterexample** (an example that disproves the statement). The statement is false. Divisibility by 5 is not sufficient for ending in zero.

(b) The fact that a number is divisible by 5 is implied by its ending in 0. The statement is true. Divisibility by 5 is a necessary condition for ending in 0.

(c) Angles equal is implied by sides equal. The statement is true.

(d) Angles equal implies sides equal. This is also true.

(c) and (d) could be replaced by 'the angles of a triangle are equal $\Leftrightarrow$ the sides of a triangle are equal', as the implication works both ways.

Conjectures, theorems and proofs

A **conjecture** is a statement that is thought to be true. It becomes a **theorem** once it is proved. A conjecture is not proved by any number of examples that confirm it. A single example that does not agree with the conjecture – a **counterexample** – disproves the conjecture.

For example, you may make the conjecture that all numbers of the form 2^k-1, where $k\geq 1$, are prime:

$k\geq 1 \Rightarrow 2^k-1$ prime

The first few examples seem to confirm the conjecture:

$2^1-1 = 1$, $2^2-1 = 3$, $2^3-1 = 7$

But this does not prove the conjecture. In fact, the next example is a counterexample: $2^4-1 = 15 = 3 \times 5$.

So, given a conjecture, how is it proved? Three methods are described below. Depending on your awarding body, you may need to know just the first method or all three.

WEB TIP
There's more about Proof by deduction on the AS Guru™ website.

Proof by deduction

Statements follow logically from one to the next until you deduce the theorem.

EXAMPLE The product of two odd numbers is odd.

Solution. Let the numbers be $2m+1$ and $2n+1$, where m and n are integers.

Their product is $(2m+1) \times (2n+1)$

$= 4mn + 2m + 2n + 1$

$= 2\ (2mn + m + n) + 1$

which is odd.

EXAMPLE The product of three consecutive numbers is divisible by 6.

SOLUTION Any two consecutive numbers must include a multiple of 2, and any three consecutive numbers must include a multiple of 3. Therefore the product of three consecutive numbers must be a multiple of both 2 and of 3, hence of $2 \times 3 = 6$.

EXAMPLE $1 + 2 + 3 + \ldots + (n-1) + n = \frac{1}{2}n(n+1)$

SOLUTION Let $S = 1 + 2 + \ldots + (n-1) + n.$

Then $S = n + (n-1) + \ldots + 2 + 1$ (reversing the order)

So $2S = (n+1) + (n+1) + \ldots + (n+1) + (n+1)$

So $2S = n \times (n+1)$

$\Rightarrow S = \frac{1}{2}n(n+1)$

These questions often involve trigonometrical identities.

EXAMPLE Show that

$$\frac{1}{\cos^2\theta} = 1 + \tan^2\theta$$

SOLUTION Starting with one side, show that it is the same as the other.

$1 + \tan^2\theta = 1 + \frac{\sin^2\theta}{\cos^2\theta}$ (as $\frac{\sin\theta}{\cos\theta} = \tan\theta$)

$= \frac{\cos^2\theta + \sin^2\theta}{\cos^2\theta}$

$= \frac{1}{\cos^2\theta}$ (as $\sin^2\theta + \cos^2\theta = 1$)

GURU TIP
See page 28, trigonometric ratios and functions.

Proof by exhaustion

Test all possible cases. If you find no counterexample then the statement must be true.

EXAMPLE No square number ends in a 2.

SOLUTION As it is only the last digit of the number you are squaring that will affect the last digit of the answer, proof by exhaustion won't be too exhausting!

As the product of two odd numbers is odd (see above), we don't need to test 1, 3, 5, 7 or 9.

Now:

$0^2 = 0$

$2^2 = 4$

$4^2 = 16$

$6^2 = 36$

$8^2 = 64$

Having examined all possibilities without finding a counterexample, the statement must be true.

GURU TIP
This can be like using a sledgehammer to crack a nut. Only use this method if there are a small number of possibilities!

WEB TIP
There's more about Proof by exhaustion and Proof by contradiction on the AS Guru™ website.

Proof by contradiction

Assume that the theorem is false; then try to deduce a contradiction.

EXAMPLE Show that $\sqrt{2}$ is irrational.

SOLUTION Assume that $\sqrt{2}$ is rational. Suppose $\sqrt{2} = \frac{a}{b}$ where a and b are integers with no common factor.

$\sqrt{2} = \frac{a}{b} \Rightarrow 2 = \frac{a^2}{b^2} \Rightarrow 2b^2 = a^2$ (1)

$\Rightarrow a^2$ is even $\Rightarrow a$ is even (as the product of two odd numbers is odd)

As a is even we can write $a = 2m$ where m is an integer. Substituting into (1) gives

$2b^2 = 4m^2 \Rightarrow b^2 = 2m^2$

$\Rightarrow b^2$ is even

$\Rightarrow b$ is even

So a and b are both even. This contradicts our assumption that they have no common factor. Therefore $\sqrt{2}$ is irrational.

KEY SKILLS
In pure mathematics you will have plenty of opportunity to develop skills and generate evidence for the Key Skill Application of number, Level 3.

Indices and surds

Algebraic expressions

When manipulating algebraic expressions such as $x^5 + 3x - 2$, there are a few rules we need to know – and a bit of terminology.

Each part of the expression is called a **term**. So in our example, $4x^5$, $3x$ and -2 are all terms.

The number in front of a power of x is called the **coefficient** of the term. Any term that is just a number is called a **constant**.

Each term is named by its power of x.

In our example, the coefficient of the x^5 term is 4.

Multiplying two terms

To multiply two terms, multiply the coefficients and the powers of x separately. Remember that

$$x^m \times x^n = x^{m+n}$$

So when multiplying two powers of x you add the indices.

EXAMPLE Simplify $3x^3 \times 5x$.

SOLUTION First multiply the coefficients: $3 \times 5 = 15$.

Next multiply the powers: $x^3 \times x = x^3 \times x^1 = x^{3+1} = x^4$.

So $3x^3 \times 5x = 15x^4$

The rules of indices

The main rules and facts about indices are summarised below.

Multiplying: $x^m \times x^n = x^{m+n}$

Dividing: $x^m \div x^n = x^{m-n}$

The power zero: $x^0 = 1$

Raising to a power: $(x^m)^n = x^{mn}$

Negative indices: $x^{-n} = \frac{1}{x^n}$

Fractional indices: $x^{\frac{m}{n}} = \sqrt[n]{x^m}$

GURU TIP
Whenever you can, take the root first.

Examples

(a) $x^3 \times x^4 = x^{3+4} = x^7$

(b) $x^5 \times x = x^{5+1} = x^6$

(c) $3x^2 \times 2x^4 = (3 \times 2)\ x^{2+4} = 6x^6$

(d) $x^4 \div x^3 = x^{4-3} = x^1 = x$

(e) $12x^5 \div 3x^3 = (12 \div 3)\ x^{5-3} = 4x^2$

(f) $x^3 \times \frac{1}{x^2} = x^3 \times x^{-2} = x^1 = x$

or $x^3 \times \frac{1}{x^2} = x^3 \div x^2 = x^1 = x$

(g) $(x^2)^3 = x^{2\times3} = x^6$

(h) $(4x^2)^3 = 4^3x^{2\times3} = 64x^6$

(i) $(x^{-2})^3 = x^{-2\times3} = x^{-6} = \frac{1}{x^6}$

(j) $x \times x^{\frac{1}{2}} = x^{1+\frac{1}{2}} = x^{\frac{3}{2}}$

(k) $27^{\frac{2}{3}} = (\sqrt[3]{27})^2 = 3^2 = 9$

(l) $8^{\frac{5}{3}} = (\sqrt[3]{8})^5 = 2^5 = 32$

Surds

Many calculations will involve finding roots of numbers.

EXAMPLE Find the length of the diagonal of a unit square.

SOLUTION Use Pythagoras' Theorem:

Length of hypotenuse = $\sqrt{(1^2+1^2)} = \sqrt{2} = 1.41...$

You may remember that $\sqrt{2}$ is an irrational number, that is, it cannot be expressed in the form $\frac{a}{b}$ where a and b are integers. This implies that any decimal representation is an approximation. The answer $\sqrt{2}$ is exact, whereas 1.41 is an approximation. '$\sqrt{2}$' is also neater. Unless you are asked to give an answer to a number of decimal places or significant figures, leave it in terms of roots. This is called **surd** notation.

You may be thinking, 'what about the other solution?' This is ignored; $\sqrt{2}$ is defined as the *positive* square root of 2.

A surd is a number of the form $\sqrt{a}$ where a is a positive integer.

As $\sqrt{a} = a^{\frac{1}{2}}$ we can obtain the following rules for manipulating surds:

$$\sqrt{x} \times \sqrt{y} = \sqrt{(xy)}$$
$$\sqrt{x} \div \sqrt{y} = \sqrt{(\tfrac{x}{y})}$$

Examples

(a) $\sqrt{2} \times \sqrt{2} = 2$

(b) $\sqrt{3} \times \sqrt{2} = \sqrt{(3\times2)} = \sqrt{6}$

(c) $\sqrt{8} = \sqrt{(4\times2)} = \sqrt{4} \times \sqrt{2} = 2\sqrt{2}$
(look for factors that are square numbers and 'bring out through the root')

(d) $\sqrt{3} \times \sqrt{8} = \sqrt{24} = \sqrt{(4\times6)} = \sqrt{4}\sqrt{6} = 2\sqrt{6}$

(e) $\sqrt{3} \div \sqrt{2} = \sqrt{\frac{3}{2}}$

(f) $\sqrt{12} + \sqrt{27} = \sqrt{(4\times3)} + \sqrt{(9\times3)} = 2\sqrt{3} + 3\sqrt{3} = 5\sqrt{3}$ (collecting together like terms)

WEB TIP
There's more about Surds on the AS Guru™ website.

Pure

Rationalising the denominator

A fraction with a surd in the denominator can be manipulated to obtain an integer denominator. Multiply the top and bottom of the fraction by the surd. This process is called 'rationalising the denominator'.

Examples

(a) $\frac{2}{\sqrt{3}} = \frac{2}{\sqrt{3}} \times \frac{\sqrt{3}}{\sqrt{3}} = \frac{2\sqrt{3}}{\sqrt{3}\sqrt{3}} = \frac{2\sqrt{3}}{3}$

(b) $\frac{\sqrt{3}}{2\sqrt{2}} = \frac{\sqrt{3}}{2\sqrt{2}} \times \frac{\sqrt{2}}{\sqrt{2}} = \frac{\sqrt{3}\sqrt{2}}{2\sqrt{2}\sqrt{2}} = \frac{\sqrt{6}}{2\times2} = \frac{\sqrt{6}}{4}$

If the denominator includes a number and a surd, multiply top and bottom by the denominator with the sign of the surd reversed:

$$\frac{1}{3-2\sqrt{2}} = \frac{1}{3-2\sqrt{2}} \times \frac{3+2\sqrt{2}}{3+2\sqrt{2}} = \frac{3+2\sqrt{2}}{9-8} = 3+2\sqrt{2}$$

GURU TIP
This works because $(a+b)(a-b) = a^2-b^2$

The equation of a straight line

The general equation of a straight line is

$y = mx + c$

where

- m is the gradient (steepness) of the line
- c is the intercept on the y axis – so the graph passes through the point $(0,c)$

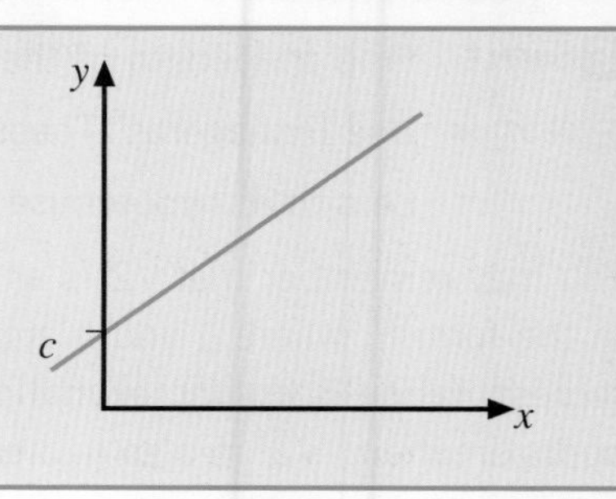

The gradient of a straight line passing through two points

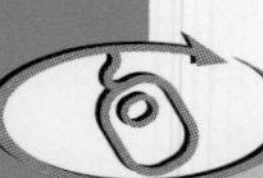

WEB TIP
There's more about Straight line graphs on the AS Guru™ website.

The gradient of the line joining (x_1,y_1) and (x_2,y_2) is $m = \frac{y_2-y_1}{x_2-x_1}$

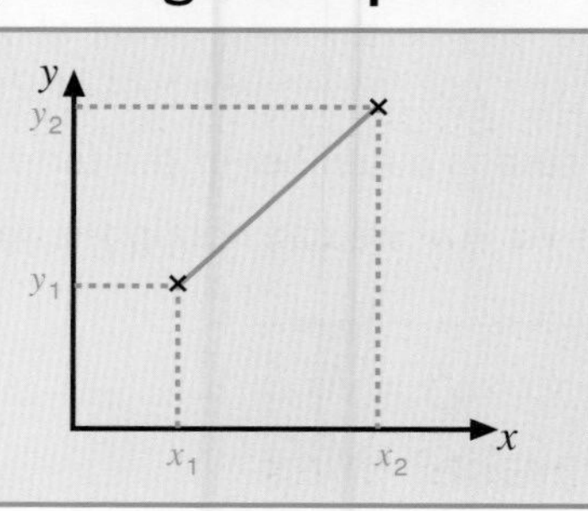

EXAMPLE Find the gradient of the straight line passing through the points (1,0) and (3,–1).

SOLUTION Use the equation above. take care to be consistent with the indices.

Taking (3,–1) as (x_2,y_2),

$m = \frac{-1-0}{3-1} = -\frac{1}{2}$

The midpoint of a line joining two points

The coordinates of the midpoint are the averages of the coordinates of the endpoints.

EXAMPLE Find the midpoint of (2,3) and (8,–3).

SOLUTION $\frac{2+8}{2} = 5$

$\frac{3+-3}{2} = 0$

So the midpoint is (5,0).

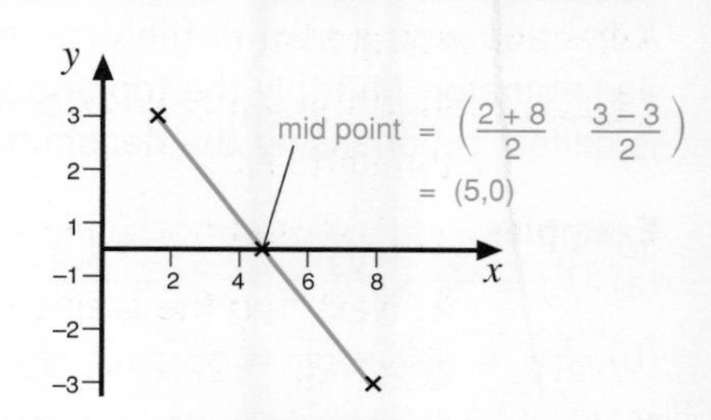

Finding the gradient of a line not given in standard form

EXAMPLE Find the gradient of the line $3y + 2x - 1 = 0$.

SOLUTION Rearrange to get the standard form $y = mx + c$:

$3y + 2x - 1 = 0$

$\Rightarrow 3y = -2x + 1$

$\Rightarrow y = -\frac{2}{3}x + \frac{1}{2}$

The gradient of the line is $-\frac{2}{3}$.

Finding the equation given the gradient and a point

> The equation of a line with gradient m passing through (x_1,y_1) is
> $y-y_1 = m(x-x_1)$

EXAMPLE Find the equation of the line with gradient $\frac{1}{2}$ which passes through the point (3,1).

SOLUTION Use the formula $y-y_1 = m(x-x_1)$. The equation is:

$y-1 = \frac{1}{2}(x-3)$

$\Rightarrow 2y-2 = x-3$

$\Rightarrow 2y = x-2$

Pure

The equation of a line passing through two points

> The equation of a line passing through (x_1,y_1) and (x_2,y_2) is
> $\frac{y-y_1}{y_2-y_1} = \frac{x-x_1}{x_2-x_1}$

GURU TIP
Multiply through to remove any fractions.

Parallel lines

Parallel lines have the same gradient.

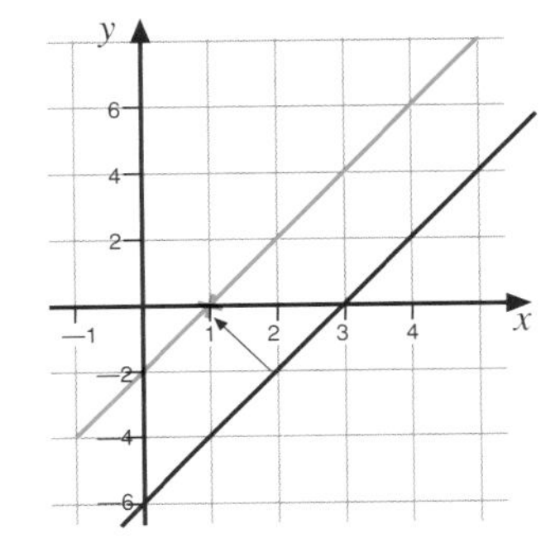

EXAMPLE Find the equation of the line that passes through (1,0) and is parallel to $y - 2x + 3 = 0$.

SOLUTION 1. First find the gradient m of the line $y - 2x + 3 = 0$:

$y - 2x + 3 = 0 \Rightarrow y = 2x - 3$, so $m = 2$.

2. Now given the gradient and a point you can find the equation using the formula $y-y_1 = m(x-x_1)$:

$y-0 = 2(x-1)$

$\Rightarrow y = 2x - 2$

Perpendicular lines

Two lines with gradients m_1 and m_2 are perpendicular $\Leftrightarrow m_1 \times m_2 = -1$.

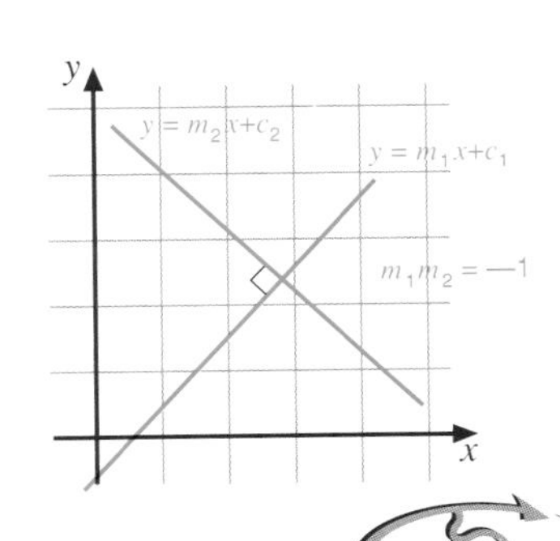

EXAMPLE Find the equation of the line perpendicular to $2y = x + \frac{1}{2}$ and passing through (1,2).

SOLUTION 1. First find the gradient of the given line: $m_1 = \frac{1}{2}$ (by inspection).

2. Next find the gradient of the perpendicular:
$\frac{1}{2} \times m_2 = -1 \Rightarrow m_2 = -2$.

3. Given the gradient and a point you can now find the equation of the line:

$y-2 = -2(x-1)$

$\Rightarrow y = -2x + 4$

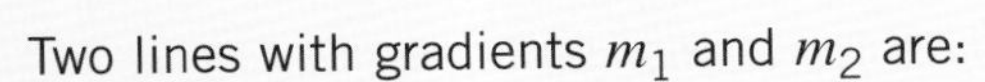

> Two lines with gradients m_1 and m_2 are:
> - parallel $\Leftrightarrow m_1 = m_2$
> - perpendicular $\Leftrightarrow m_1m_2 = -1$

WEB TIP
There's more about Parallel and perpendicular lines on the AS Guru™ website.

Polynomials 1 – working with polynomials

An expression such as $y = 3x^3 - 10x^2 - 6$, which is made up of multiples of non-negative whole-number powers of x, is called a **polynomial**.

(Note that a polynomial contains only non-negative whole-number powers of x, so terms like $\frac{3}{x^2}$ or $4x^{\frac{2}{3}}$ are not allowed.)

The highest power of x in the example above is x^3. The polynomial is said to be of third degree. The **degree** of a polynomial tells you the highest power of x.

It is normal to write polynomials in order of decreasing powers.

The number in front of each power of x is called the **coefficient** of that power. So in our example, -10 is the coefficient of x^2.

Remember: although most polynomials you see will be in x, they can be in any symbol. In mechanics you will meet a lot of of polynomials in t.

WEB TIP
There's more about Algebraic manipulation on the AS Guru™ website.

Adding and subtracting polynomials

You will remember from GCSE that you can only add or subtract like terms. You can simplify $2x^3 + 5x^3$ to $7x^3$. You can't simplify $2x^3 + 5x^2$.

EXAMPLE Let $A = 2x^2 - 6x + 10$, $B = 4x^3 - 2x + 1$, $C = 2x^2 - 2x + 10$.

Write the following as polynomials in x. State the degree of the polynomials.

(a) $A + B$

(b) $A - C$

SOLUTION (a) $A + B = (2x^2 - 6x + 10) + (4x^3 - 2x + 1)$

$= 2x^2 + 4x^3 - 6x - 2x + 10 + 1$ (collecting together like terms)

$= 4x^3 + 2x^2 - 8x + 11$ (adding like terms)

The polynomial is of degree 3.

(b) $A - C = (2x^2 - 6x + 10) - (2x^2 - 2x + 10)$

$= 2x^2 - 6x + 10 - 2x^2 + 2x - 10$ (remember the minus sign outside the bracket changes all the signs inside the bracket)

$= 2x^2 - 2x^2 - 6x + 2x + 10 - 10$

$= -4x$

The polynomial is of degree 1.

Multiplication of polynomials

You have done this at GCSE when multiplying out brackets.

EXAMPLE Find $A \times A$.

SOLUTION $(2x^2 - 6x + 10) \times (2x^2 - 6x + 10) = 2x^2 \times (2x^2 - 6x + 10) - 6x \times (2x^2 - 6x + 10) + 10 \times (2x^2 - 6x + 10)$

$= 4x^4 - 12x^3 + 20x^2 - 12x^3 + 36x^2 - 60x + 20x^2 - 60x + 100$

$= 4x^4 - 12x^3 - 12x^3 + 20x^2 + 36x^2 + 20x^2 - 60x - 60x + 100$

$= 4x^4 - 24x^3 + 76x^2 - 120x + 100$ (a polynomial of degree 4)

Division by inspection

If you know the division has no remainder, use 'unmultiplication'.

EXAMPLE Divide $x^3 + 5x^2 + 10x + 8$ by $x + 2$, given that $x + 2$ is a factor.

SOLUTION To get the x^3 term when multiplying by $x + 2$ you need an x^2 term. Also, to get the constant term 8 when multiplying by $x + 2$ you need a constant term 4. So the polynomial must be of the form $x^2 + Ax + 4$.

Multiplying:

$(x + 2) \times (x^2 + Ax + 4) = x^3 + Ax^2 + 4x + 2x^2 + 2Ax + 8$

Comparing the coefficient of x^2 with that of $x^3 + 5x^2 + 10x + 8$ gives

$A + 2 = 5 \Rightarrow A = 3$

We can check this by looking at the coefficients of x:

$4 + 2A = 10$ ✓

So $x^3 + 5x^2 + 10x + 8 = (x + 2)(x^2 + 3x + 4)$

Pure

Long division

You may prefer this method, which you will need to use if the expression you are dividing by is not a factor.

GURU TIP
When possible, use division by inspection. Long division takes longer!

EXAMPLE Divide $2x^3 - 3x^2 - 100$ by $x - 4$. What is the remainder?

SOLUTION As we want a $2x^3$ term when we multiply by $x - 4$, the x^2 term must be $2x^2$.

Multiplying $2x^2 \times (x - 4)$ gives a $-8x^2$ term. As we want $-3x^2$, the x term must be $+5x$.

Multiplying $5x \times (x - 4)$ gives a $-20x$ term. As we want $0x$, the constant term must be $+20$.

Multiplying $20 \times (x - 4)$ gives a -80 constant term. As we want -100, the remainder of the division is -20.

So $2x^3 - 3x^2 - 100 = (x - 4)(2x^2 + 5x + 20) - 20$

$$
\begin{array}{r|l}
 & 2x^2 + 5x + 20 \\
x - 4 & 2x^3 - 3x^2 + 0x - 100 \\
 & -\ \underline{2x^3 - 8x^2} \\
 & 5x^2 \\
 & -\ \underline{5x^2 - 20x} \\
 & 20x - 100 \\
 & -\ \underline{20x - 80} \\
 & -20
\end{array}
$$

GURU TIP
Set out your answer like this.

Polynomials 2 – factorising

Roots of quadratic equations

The **roots** of an equation $f(x) = 0$ are the points where the graph of $y = f(x)$ cuts the x axis. That is, they are the solutions of the equation $f(x) = 0$.

Let's start with polynomials of degree 2, which you probably know as quadratic equations.

EXAMPLE Find the roots of $x^2 - 5x + 6 = 0$.

SOLUTION Factorise. We need to think of a pair of numbers whose *product* is +6 and whose *sum* is –5.

Trial and error yields the numbers –3 and –2.

So $x^2 - 5x + 6 = (x - 2)(x - 3)$.

So we need to solve $(x - 2)(x - 3) = 0$.

For this to be true, *either* $(x - 2) = 0$ ($\Rightarrow x = 2$) *or* $(x - 3) = 0$ ($\Rightarrow x = 3$).

So the two roots of the equation are 2 and 3.

The above process should be second nature to you. If it isn't, practise it!

Let's summarise what we know about quadratic equations. If $f(x)$ is a polynomial of degree 2, then:

- $f(x) = 0$ has at most 2 solutions – so the graph crosses the x axis at most twice
- if $(x - a)$ is a factor of $f(x)$ then a is a root of the equation $f(x) = 0$

In general:

> If $f(x)$ is a polynomial of degree n, then $f(x) = 0$ has at most n solutions. The graph crosses the x axis at most n times.

GURU TIP
Make sure you know how to apply this theorem – you will almost certainly have to in your exam.

The Factor Theorem

Let $f(x)$ be a polynomial. If $(x - a)$ is a factor of $f(x)$ then a is a root of the equation $f(x) = 0$.

The converse of this is also true. That is, if a is a root of the equation $f(x) = 0$, then $(x - a)$ is a factor of $f(x)$.

> $(x - a)$ is a factor of $f(x) \Leftrightarrow f(a) = 0$
>
> More generally,
>
> $(bx - a)$ is a factor of $f(x) \Leftrightarrow f(\frac{a}{b}) = 0$

WEB TIP
There's more about The Factor Theorem on the AS Guru™ website.

If you are given one of the factors of a polynomial, you can divide by that factor (see page 15) and so factorise the polynomial. But how do you make a start if you are not given a factor?

EXAMPLE Factorise $f(x) = x^3 + 2x^2 - x - 2$

SOLUTION 1. We haven't got much to go on, but the only (integer) factors of the constant term -2 are ± 1 and ± 2. This tells us that any factor must be of the form $(x \pm 1)$ or $(x \pm 2)$.

But how do we tell which? This is where the Factor Theorem comes in.

2. Evaluate $f(1) = 1^3 + 2 \times 1^2 - 1 - 2 = 0$.

So by the Factor Theorem, $(x - 1)$ is a factor of $f(x)$.

3. Next divide the function by the factor. By inspection we need a polynomial of the form $x^2 + Ax + 2$.

Multiplying: $(x - 1) \times (x^2 + Ax + 2) = x^3 + Ax^2 + 2x - x^2 - Ax - 2$

Comparing coefficients of x^2:

$A - 1 = 2 \Rightarrow A = 3$

Check by comparing coefficients of x:

$2 - A = -1$ ✓

So $f(x) = (x - 1)(x^2 + 3x + 2)$

4. Finally, factorise the resulting quadratic expression (if possible):

$x^2 + 3x + 2 = (x + 1)(x + 2)$

So $f(x) = (x - 1)(x + 1)(x + 2)$

Note the alternative form of the theorem:

$(x + a)$ is a factor of $f(x) \Leftrightarrow f(-a) = 0$

$(bx + a)$ is a factor of $f(x) \Leftrightarrow f(-\frac{a}{b}) = 0$

Pure

The binomial expansion

Expansion of $(1 + x)^n$

We often need to expand expressions like $(1 + x)^2$.

$(1 + x)^2 = (1 + x)(1 + x) = 1 + x + x + x^2 = 1 + 2x + x^2$

More generally if you raise $(1 + x)$ to the power of n you get a binomial expansion.

There is a pattern to the results:

$$(1 + x)^0 = 1$$
$$(1 + x)^1 = 1 + x$$
$$(1 + x)^2 = 1 + 2x + x^2$$
$$(1 + x)^3 = 1 + 3x + 3x^2 + x^3$$
$$(1 + x)^4 = 1 + 4x + 6x^2 + 4x^3 + x^4$$

The binomial coefficients

The binomial coefficients are given by Pascal's triangle.

GURU TIP
Learn how to generate this triangle.

1

1 1

1 2 1

1 3 3 1

1 4 6 4 1

1 5 10 10 5 1

1 6 15 20 15 6 1

Each row begins with and ends with a 1. To find any other entry, add the two numbers immediately above. For example, the 15 in the bottom row is obtained by adding 5 and 10.

However, you don't have to write out Pascal's triangle every time you want to calculate a binomial coefficient.

The binomial coefficients are given by

$${}^nC_r = \frac{n!}{r!(n-r)!}$$

where n is the power of the expansion and r is the power of the term.

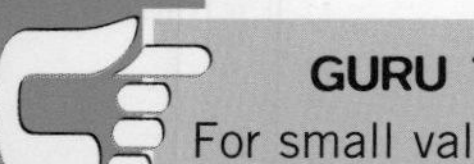

GURU TIP
For small values of n, it is easier to use the triangle than the formula.

For example, if you wanted the coefficient of x^5 in the expansion of $(1 + x)^{20}$, you would calculate:

$${}^{20}C_5 = \frac{20!}{5!(20-5)!} = \frac{20!}{5!15!} = \frac{20\times19\times18\times17\times16}{5\times4\times3\times2\times1} = 15504.$$

If you are doing a statistics module including the binomial distribution, you may recognise this as the formula for the number of combinations of r objects chosen from n.

Note: The first entry in any row of the triangle of coefficients is ${}^nC_0 = 1$.

Expansion of a general binomial expression

We can use Pascal's triangle to expand $(x + y)$ to different powers:

$$(x+y)^0 = 1$$
$$(x+y)^1 = x + y$$
$$(x+y)^2 = x^2 + 2xy + y^2$$
$$(x+y)^3 = x^3 + 3x^2y + 3xy^2 + y^3$$

EXAMPLE Expand $(2a - b)^3$.

SOLUTION Use Pascal's triangle to find the required expansion of $(x + y)^3$, and substitute $x = 2a$, $y = -b$:

$(2a)^3 + 3(2a)^2(-b) + 3(2a)(-b)^2 + (-b)^3 = 8a^3 - 12a^2b + 6ab^2 - b^3$

Using the binomial theorem in calculations

EXAMPLE Find the binomial expansion of $(1 + x)^6$. Use this to find 0.99^6 to 5 d.p.

SOLUTION Using Pascal's triangle:

$(1 + x)^6 = 1 + 6x + 15x^2 + 20x^3 + 15x^4 + 6x^5 + x^6$

Now $0.99 = 1 + (-0.01)$. So

$0.99^6 = 1 + 6(-0.01) + 15(-0.01)^2 + 20(-0.01)^3 + 15(-0.01)^4 + \dots$

$= 1 - 0.06 + 15 \times 0.0001 - 20 \times 0.000001 + 15 \times 0.00000001$
(don't include any more terms as they are too small)

$= 1 - 0.06 + 0.0015 - 0.00002 + 0.00000015$

$= 0.94148$ to 5 d.p.

GURU TIP
If you get a question like this in your exam, check your result using a calculator.

The link with the binomial distribution

The possible outcomes when tossing a coin 1, 2 or 3 times are listed below:

1: H or T

2: HH or HT/TH or TT

3: HHH or HHT/HTH/THH or TTH/THT/HTT or TTT

Can you see Pascal's triangle here?

The number of ways of getting exactly r heads when tossing n coins is ${}^nC_r$.

Pure

Solving quadratic equations

WEB TIP
There's more about Completing the square on the AS Guru™ website.

Completing the square

When trying to factorise quadratics you sometimes feel that the quadratic *almost* factorises!

EXAMPLE Solve $x^2 - 6x + 7 = 0$.

SOLUTION If this were $x^2 - 6x + 9$, it would factorise as $(x - 3)^2$.

$x^2 - 6x + 7 = (x - 3)^2 - 2$

This is called 'completing the square'.

$(x - 3)^2 - 2 = 0 \Leftrightarrow (x - 3)^2 = 2$

$\Leftrightarrow x - 3 = \pm\sqrt{2}$

$\Leftrightarrow x = 3 \pm \sqrt{2}$

Here's how to complete the square in general. First, let's assume that the coefficient of x^2 is 1.

1. Halve the coefficient of the x term. This number goes with the x in the bracket.
2. Square and compare.

Examples

(a) $f(x) = x^2 + 6x + 7$.

1. Half of 6 is 3, so try $(x + 3)^2$
2. $(x + 3)^2 = x^2 + 6x + 9 = f(x) + 2$

So $f(x) = (x + 3)^2 - 2$

(b) $f(x) = x^2 - 9x + 20$.

1. Half of -9 is $-\frac{9}{2}$, so try $(x - \frac{9}{2})^2$
2. $(x - \frac{9}{2})^2 = x^2 - 9x + \frac{81}{4} = f(x) + \frac{1}{4}$

So $f(x) = (x - \frac{9}{2})^2 - \frac{1}{4}$

If the coefficient of x^2 is not 1, take the coefficient of x^2 out of the x^2 and x terms as a factor, and complete the square in the bracket as above.

Examples

(a) $f(x) = 3x^2 + 6x + 7$

$f(x) = 3(x^2 + 2x) + 7$

1. Half of 2 is 1, so try $(x + 1)^2$
2. $3(x + 1)^2 = 3x^2 + 6x + 3 = f(x) - 4$

So $f(x) = 3(x + 1)^2 + 4$

(b) $f(x) = 2x^2 - 8x + 18$

$f(x) = 2(x^2 - 4x) + 18$

1. Half of -4 is -2, so try $(x - 2)^2$
2. $2(x - 2)^2 = 2x^2 - 8x + 8 = f(x) - 10$

So $f(x) = 2(x - 2)^2 + 10$

Vertex and line of symmetry

Once you have a quadratic function $f(x)$ in completed-square form you can locate the vertex and line of symmetry of the graph of $y = f(x)$ without plotting. This is useful when you are asked to sketch a quadratic.

EXAMPLE Find the vertex and line of symmetry of the graph of $y = 2x^2 - 8x + 18$.

SOLUTION Completing the square:

$y = 2(x - 2)^2 + 10$ (see above)

The minimum value of y is 10 and this occurs when $x - 2 = 0$.

The vertex is $(2,10)$ and the line of symmetry is $x - 2 = 0$ or $x = 2$.

The quadratic formula

By completing the square of a general quadratic, you get the formula for its roots:

> The roots of the quadratic equation $ax^2 + bx + c = 0$ are given by
> $x = \frac{-b \pm \sqrt{(b^2 - 4ac)}}{2a}$

EXAMPLE Solve the equation $x^2 - 6x + 7 = 0$, giving your answer to 3 d.p.

SOLUTION Use the quadratic formula with $a = 1$, $b = -6$, $c = 7$:

$$x = \frac{-(-6) \pm \sqrt{((-6)^2 - 4 \times 1 \times 7)}}{2 \times 1}$$
$$= \frac{6 \pm \sqrt{8}}{2}$$
$$= \frac{6 \pm 2\sqrt{2}}{2}$$
$$= 3 \pm \sqrt{2}$$

So to 3 decimal places the roots of the equation are 4.414 and 1.586.

Pure

The discriminant and the number of roots

The part of the formula under the square root is called the **discriminant**.

Depending on the values of a, b and c, this can be positive, negative or zero. The three cases are summarised in the following table.

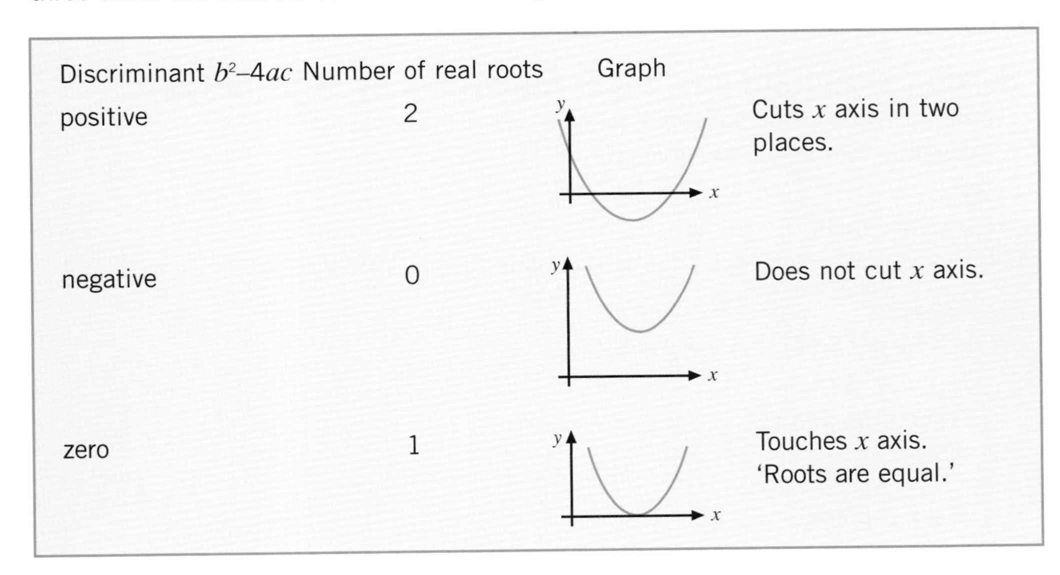

Discriminant b^2-4ac	Number of real roots	Graph	
positive	2		Cuts x axis in two places.
negative	0		Does not cut x axis.
zero	1		Touches x axis. 'Roots are equal.'

WEB TIP
There's more about Quadratic solutions on the AS Guru™ website.

Quadratics in a function of x

EXAMPLE Solve $x^{\frac{2}{3}} - 5x^{\frac{1}{3}} + 4 = 0$.

SOLUTION This is a quadratic in $x^{\frac{1}{3}}$. Make the substitution $y = x^{\frac{1}{3}}$.

Solve for y: $y^2 - 5y + 4 = 0 \Leftrightarrow (y - 4)(y - 1) = 0$

$\Leftrightarrow y = 4$ or 1

Back-substitute to find x:

$y = x^{\frac{1}{3}} \Rightarrow x = y^3 = 1$ or 64.

GURU TIP
Don't forget to back-substitute at the end.

Simultaneous equations

Simultaneous linear equations occur frequently. You need to be able to solve them by the following methods.

Solving simultaneous linear equations by elimination

1. Multiply one or both of the equations by a constant so that one of the coefficients is the same in both equations.
2. Subtract one from the other (or add if the signs are different) to eliminate one unknown.
3. Solve for the other unknown.
4. Back-substitute.
5. Check your values by substituting them into the equation you *didn't* use in Step 4.

GURU TIP
Always take the time to check your answers!

Examples

(a) $2x + 5y = 18$ (1) and $x + y = 6$ (2)

1. $(1) \Rightarrow 2x + 5y = 18$
   $2 \times (2) \Rightarrow 2x + 2y = 12$ (3)
2. $(1) - (3) \Rightarrow 3y = 6$
3. $\Rightarrow y = 2$
4. Subs. in (2) $\Rightarrow x + 2 = 6$
   $\Rightarrow x = 4.$
5. Check: (1) $\Rightarrow 2\times4 + 5\times2 = 18$ ✓

(b) (1) and $2x + y = 9$ (2)

1. $(1) \Rightarrow 2x - 3y = 13$
   $3 \times (2) \Rightarrow 6x + 3y = 27$ (3)
2. $(1) + (3) \Rightarrow 8x = 40$
3. $\Rightarrow x = 5$
4. Subs. in (2) $\Rightarrow 10 + y = 9$
   $\Rightarrow y = -1$
5. Check in (1) $\Rightarrow 2\times5 - 3\times(-1) = 13$ ✓

Solving simultaneous linear equations by substitution

1. Rearrange one of the equations to make x or y the subject.
2. Substitute into the other equation.
3. Solve.
4. Back-substitute.
5. Check your values by substituting them into the equation you *didn't* use in Step 4.

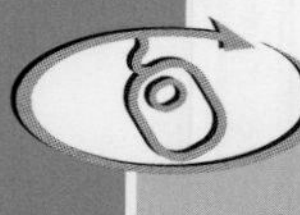

WEB TIP
There's more about Simultaneous equations on the AS Guru™ website.

Examples

(a) $2x + 5y = 18$ (1) and $x + y = 6$ (2)

1. (2) $\Rightarrow x = 6 - y$
2. Substitute into (1): $2(6 - y) + 5y = 18$
3. Solve: $12 + 3y = 18$
   $\Rightarrow y = 2.$
4. Back-substitute: (2) $\Rightarrow x + 2 = 6$
   $\Rightarrow x = 4.$

So $x = 4$ and $y = 2$.

5. Check: (1) $\Rightarrow 2\times4 + 5\times2 = 18$ ✓

(b) $2x - 3y = 13$ (1) and $2x + y = 9$ (2)

1. (2) $\Rightarrow y = 9 - 2x$
2. Substitute into (1): $2x - 3(9 - 2x) = 13$
3. Solve: $8x - 27 = 13$
   $\Rightarrow x = 5.$
4. Back-substitute: (2) $\Rightarrow 10 + y = 9$
   $\Rightarrow y = -1.$

So $x = 5$ and $y = -1$.

5. Check: (1) $\Rightarrow 2\times5 - 3\times(-1) = 13$ ✓

Graphical interpretation of simultaneous linear equations

The solution to two simultaneous linear equations gives the coordinates of the point where the lines cross. In Example (a) above, the graphs look like this:

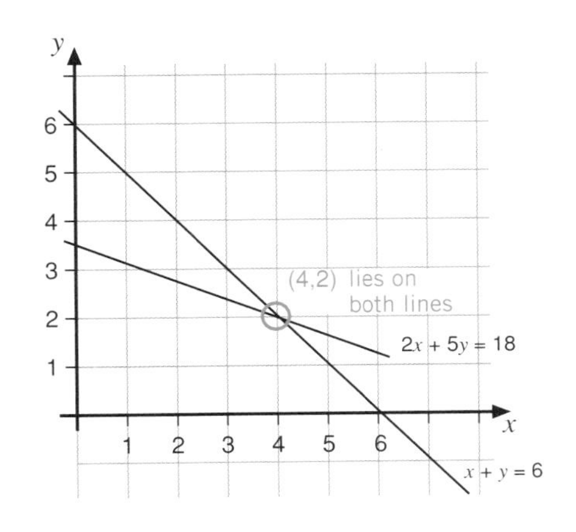

One equation linear and one nonlinear

To find the points of intersection of a line and a curve, you need to solve a linear and a nonlinear equation simultaneously. For this type of problem, use substitution.

EXAMPLE Find the points of intersection of the line $y = x + 1$ and the curve $x^2 + 2y^2 = 1$.

SOLUTION Substitute: $x^2 + 2(x + 1)^2 = 1$

$\Rightarrow x^2 + 2(x^2 + 2x + 1) = 1$

$\Rightarrow 3x^2 + 4x + 1 = 0$

$\Rightarrow (3x + 1)(x + 1) = 0$

$\Rightarrow x = -1 \text{ or } -\frac{1}{3}$.

Back-substitute into the linear equation:

$x = -1,\ y = x + 1 \Rightarrow y = 0$

$x = -\frac{1}{3},\ y = x + 1 \Rightarrow y = \frac{2}{3}$

So the points of intersection are $(-1,0)$ and $(-\frac{1}{3}, \frac{2}{3})$.

(Check these by substituting into the quadratic equation.)

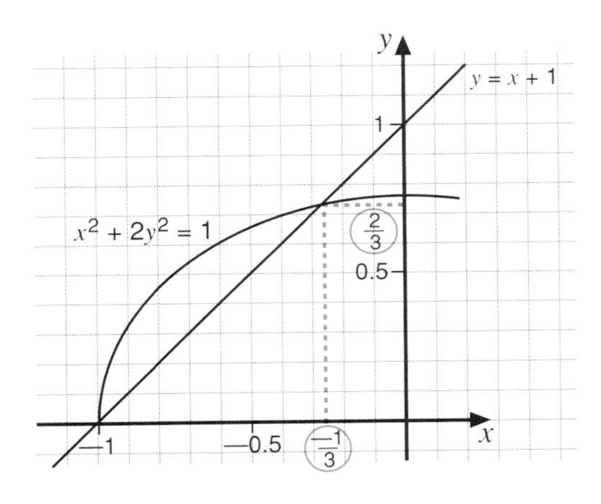

Points of intersection of two curves

EXAMPLE Find the points of intersection of the curves $x^2 + 2y^2 = 22$ and $x^2 + y = 1$.

SOLUTION First find a term that appears in both equations (x^2):

$x^2 + y = 1 \Rightarrow x^2 = 1 - y$

So $1 - y + 2y^2 = 22$

$\Rightarrow 2y^2 - y - 21 = 0$

$\Rightarrow (2y - 7)(y + 3) = 0$

$\Rightarrow y = -3 \text{ or } \frac{7}{2}$.

Back-substitute:

$y = -3,\ x^2 = 1 - y \Rightarrow x^2 = 4 \Rightarrow x = \pm 2$

$y = \frac{7}{2},\ x^2 = 1 - y \Rightarrow x^2 = -\frac{5}{2}$. *No solution.*

So the points of intersection are $(-2,-3)$ and $(2,-3)$.

(Check these by substituting into $x^2 + 2y^2 = 22$.)

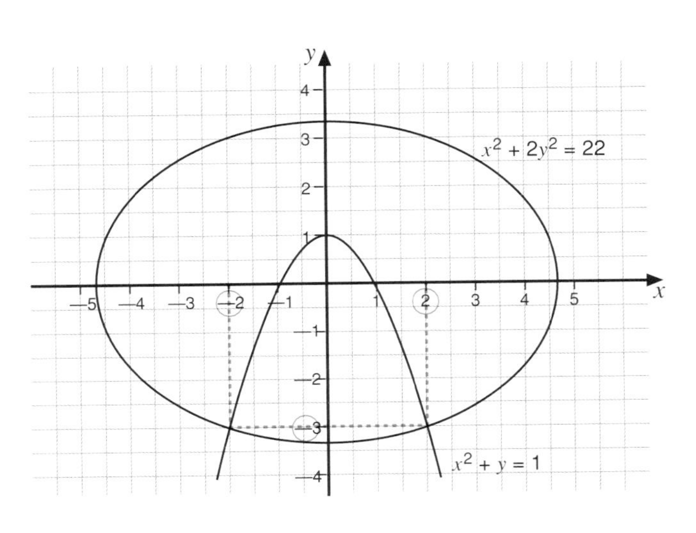

Inequalities

Algebra and notation

Working with inequalities is much like working with equations. The 'equals' sign is replaced with one of the following symbols.

$<$ 'less than'

$>$ 'greater than'

$\le$ 'less than or equal to'

$\ge$ 'greater than or equal to'

The algebra of inequalities is similar to that of equations, except:

> If you multiply or divide a linear inequality by a negative number, the direction of the inequality is reversed.

For example: $3 > 1 \Rightarrow -6 < -2$ (multiplying by –2).

Solving and representing linear inequalities

EXAMPLE Show the set of values for which:

(a) $2x + 3 \ge 7$

(b) $2(x - 6) < 3(1 - x)$

SOLUTION (a) $2x + 3 \ge 7$

$\Leftrightarrow 2x \ge 4$ (subtracting 3 from each side)

$\Leftrightarrow x \ge 2$ (dividing each side by 2)

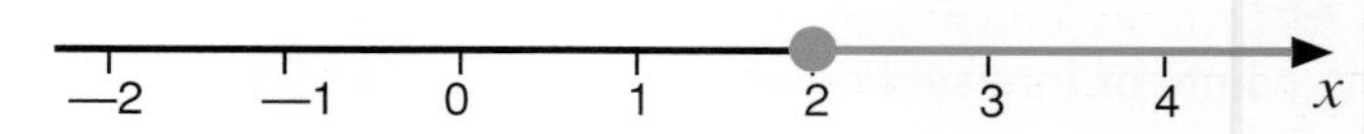

Remember: as 2 is included (greater than *or equal to*), the circle is filled in.

(b) $2(x - 6) < 3(1 - x)$

$\Leftrightarrow 2x - 12 < 3 - 3x$

$\Leftrightarrow 5x < 15$ (adding $3x + 12$ to each side)

$\Leftrightarrow x < 3$ (dividing each side by 5)

As 3 is not included, the circle is left hollow.

Quadratic inequalities

These are a bit trickier.

EXAMPLE For what values of x is $x^2 \ge 4$?

SOLUTION Taking square roots: $x \ge 2$ or $x \le -2$.

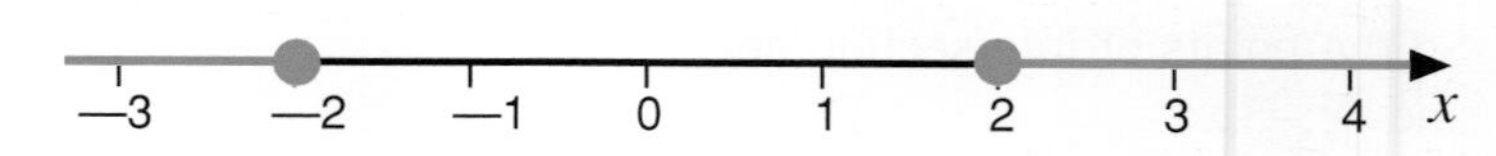

In this example, there are two sections of the number line that satisfy the inequality.

WEB TIP
There's more about inequalities and modulus on the AS Guru™ website.

EXAMPLE Solve the following inequality: $s^2 - 2s - 3 \geq 0$

(Remember, polynomials don't just come in x.)

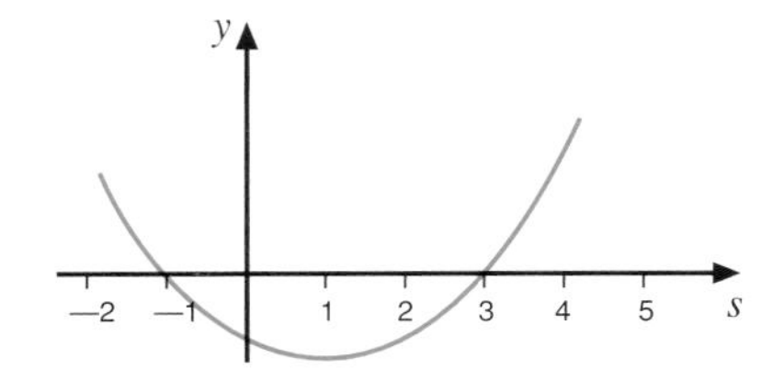

SOLUTION 1. First sketch the graph of the function. It will help to factorise it:

$s^2 - 2s - 3 \geq 0 \Leftrightarrow (s + 1)(s - 3) \geq 0$

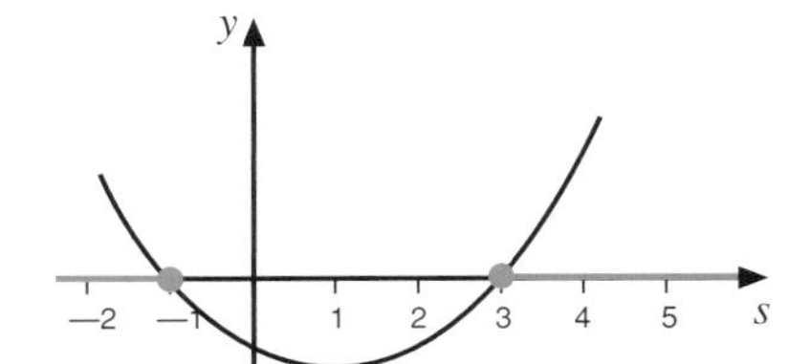

2. Now identify the sections of the s axis that satisfy the inequality. We require the two sections where $y \geq 0$. Use closed circles ($y = 0$ is included).

So the solution is: $s \geq 3$ or $s \leq -1$.

The modulus function

The modulus function tells you the size or magnitude of a number, ignoring the sign. For example, $|-5| = 5$.

$|a|$ is pronounced 'mod a'.

(Remember the modulus of a *vector* **a**, |**a**|, is the *length* of the vector.)

The graph of the modulus function looks like a V. The function is symmetrical.

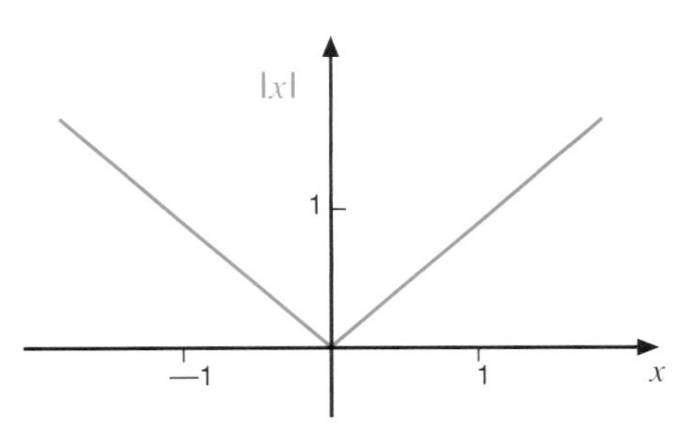

EXAMPLE Sketch the graph of $y = |2x - 3|$.

SOLUTION $2x - 3 = 0 \Leftrightarrow x = \frac{3}{2}$, so $y = 0$ at $x = \frac{3}{2}$.

So the graph is symmetrical about $(\frac{3}{2},0)$ and is linear on either side of this point. To sketch the graph you just need to work out one more point on it, say $x = 5$. When $x = 5$, $y = 7$. Draw a line from $(\frac{3}{2},0)$ through (5,7) and use symmetry to draw in the other half.

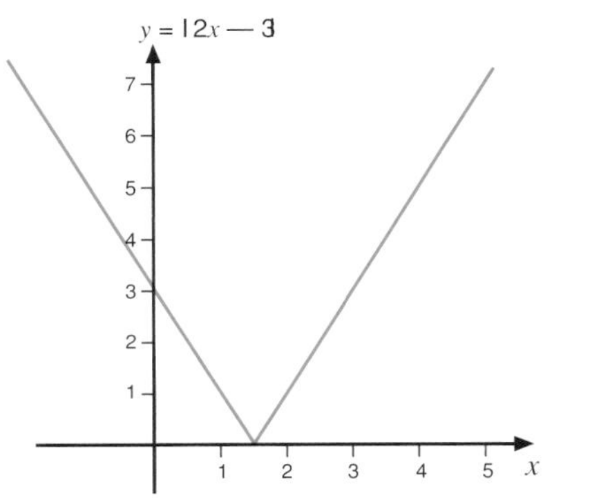

The modulus function in inequalities

EXAMPLE Solve $|x| \leq 3$.

SOLUTION Sketch the graph of $y = |x|$. Draw on the line $y = 3$.

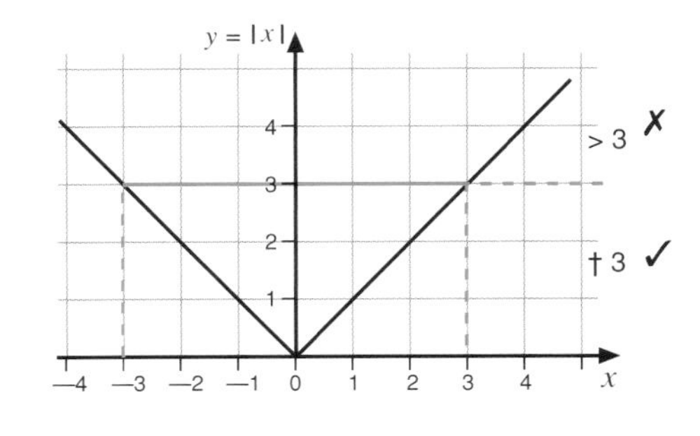

Identify the region where $y \leq 3$ (the middle section).

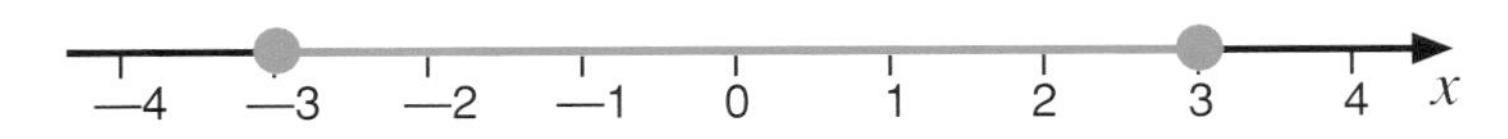

The region is $-3 \leq x \leq 3$.

Actually you don't really need to sketch the graph.

EXAMPLE $|2x - 3| < 1$.

SOLUTION If the magnitude of $2x - 3$ is less than 1 then $2x - 3$ lies between -1 and 1:

$|2x - 3| < 1 \Leftrightarrow -1 < 2x - 3 < 1$

$\Leftrightarrow 2 < 2x < 4$

$\Leftrightarrow 1 < x < 2$

Pure

Sectors and segments

Radians

You will be used to measuring angles in degrees. But it is often easier to use another unit of measurement – the **radian**.

One radian is the angle subtending an arc of the same length as the radius of the circle.

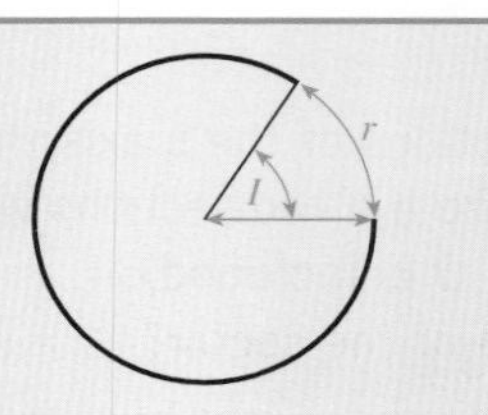

WEB TIP
There's more about Radian measure on the AS Guru™ website.

Since the circumference of a circle of radius r is $2\pi r$, a complete revolution is 2π radians.

one complete revolution = 360° = 2π radians

So 1 radian = $\frac{360°}{2\pi}$ = 57.29...°
and π radians = 180°.

Here's a quick conversion table for some commonly occurring angles:

Degrees	360	180	90	60	45	30
Radians	2π	π	$\frac{\pi}{2}$	$\frac{\pi}{3}$	$\frac{\pi}{4}$	$\frac{\pi}{6}$

In general:

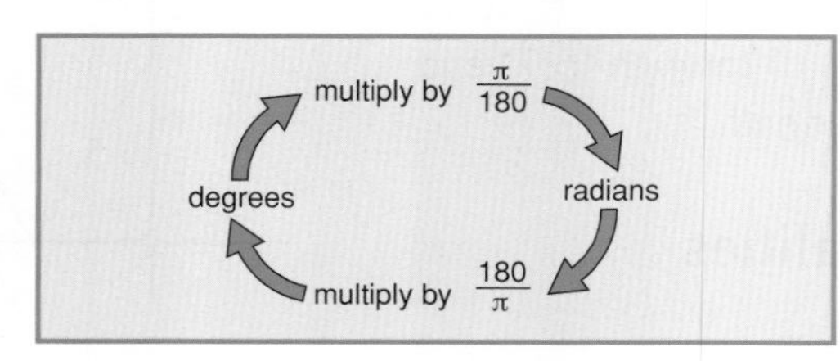

EXAMPLE Express the following angles in degrees:

(a) $\frac{\pi}{6}$ (b) $\frac{3\pi}{4}$

SOLUTION (a) $\pi = 180°$

$\Rightarrow \frac{\pi}{6} = \frac{180°}{6} = 30°$

(b) $\pi = 180°$

$\Rightarrow \frac{3\pi}{4} = \frac{3 \times 180°}{4} = 135°$

EXAMPLE Express the following angles in radians:

(a) 270° (b) 120°

SOLUTION (a) $180° = \pi$

$\Rightarrow 270° = \frac{270}{180} \times \pi = \frac{3\pi}{2}$

(b) $180° = \pi$

$\Rightarrow 120° = \frac{120}{180} \times \pi = \frac{2\pi}{3}$

Area of sector and arc length

The area of a circle of radius r is πr^2 and its circumference is $2\pi r$.

To find the area of a sector or the length of an arc subtended by a given angle, multiply the total area or length by the fraction of a complete revolution. Using radians makes the formulae very simple.

If the subtended angle is θ radians and the circle has radius r, then

arc length $s = r\theta$

area $A = \frac{1}{2} r^2\theta$

If given an angle in degrees, convert to radians and use these formulae.

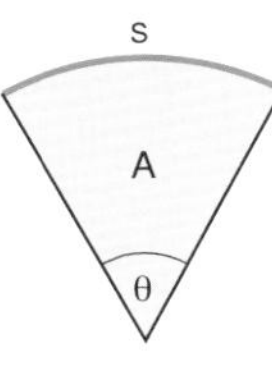

Pure

EXAMPLE A sector of a circle has arc length 10cm. If the radius of the circle is 20cm, what is the area of the sector?

SOLUTION 1. First find the angle subtended at the centre. Using $s = r\theta$:

$10 = 20\theta \Rightarrow \theta = \frac{1}{2}$ radian

2. Now find the area of the sector. Using $A = \frac{1}{2} r^2\theta$:

$A = \frac{1}{2} \times 20^2 \times \frac{1}{2} = 100$

The area of the sector is 100cm^2.

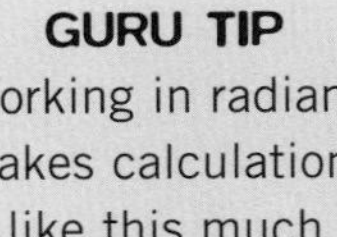

GURU TIP
Working in radians makes calculations like this much simpler.

Chords, segments and triangles

Remember the area of a triangle is $\frac{1}{2}\, ab\sin C$, where C is the angle between sides of length a and b. Applying this to the triangular region of the sector shown:

Area of triangle OAB $= \frac{1}{2} r^2 \sin\theta$

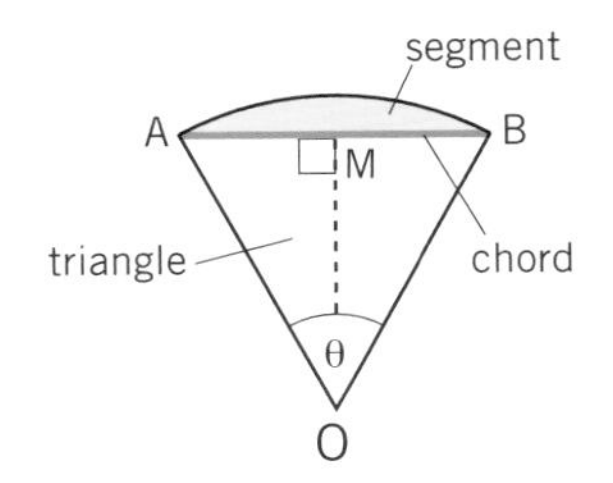

To find the area of the segment (the region between the chord AB and the arc AB), subtract the area of the triangle from the area of the sector.

(area of segment) = (area of sector) – (area of triangle OAB)

The length of the chord AB is twice the length of AM. As AMO is a right angle and $\angle AOM = \frac{1}{2} \angle AOB$:

chord length AB = 2 × AM = $2r\sin(\frac{1}{2}\theta)$

EXAMPLE Continuing with the above example, find:

(a) the length of the chord

(b) the area of the segment

SOLUTION (a) Chord length $= 2 \times 20 \times \sin(\frac{1}{2} \times \frac{1}{2}) = 40 \times \sin \frac{1}{4} = 9.90$ to 3 s.f.

The length of the chord is 9.90cm to 3 s.f.

(b) Area of triangle $= \frac{1}{2} \times 20^2 \times \sin \frac{1}{2} = 95.885$

So (area of segment) = (area of sector) – (area of triangle)

$= 100 - 95.885$

$= 4.11$

The area of the segment is 4.11cm^2 to 3 s.f.

GURU TIP
Your calculator must be in radians mode to do these calculations.

Trigonometry 1 – definitions

The trigonometric ratios

As any two right-angled triangles with an angle θ are similar to one other, the ratios of corresponding pairs of sides depend only on θ. These ratios are defined as follows:

$\sin\theta = \frac{\text{opp}}{\text{hyp}}$

$\cos\theta = \frac{\text{adj}}{\text{hyp}}$

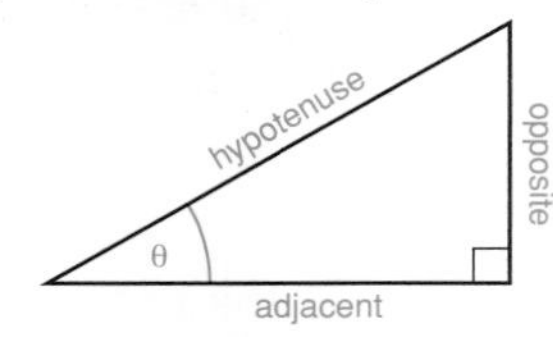

WEB TIP
There's more about Trignometric functions on the AS Guru™ website.

Trigonometric functions

Imagine an arm of length 1 sweeping out a circle.

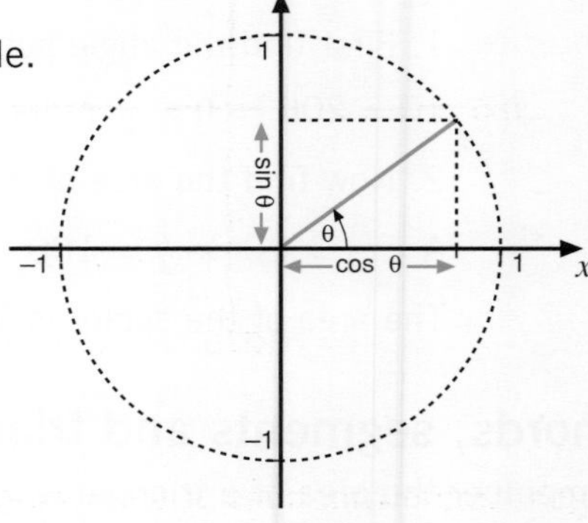

$\sin\theta = \frac{y}{1} = y$, and $\cos\theta = \frac{x}{1} = x$.

So if we plot a graph of y against θ we obtain the sine curve, and if we plot x against θ we obtain the cosine curve.

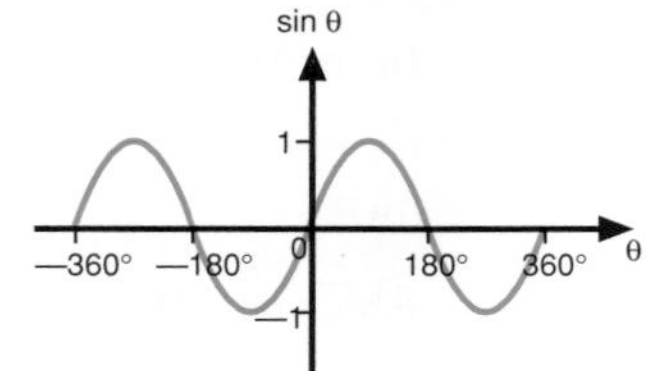

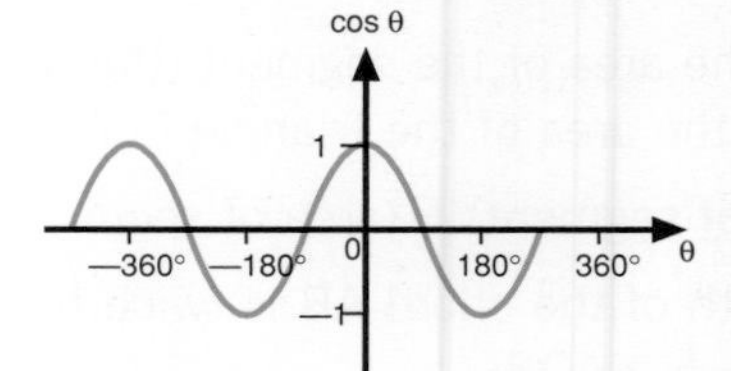

You can see these curves being generated on the website.

Notice that by applying Pythagoras' Theorem we get:

$$\sin^2\theta + \cos^2\theta = 1$$

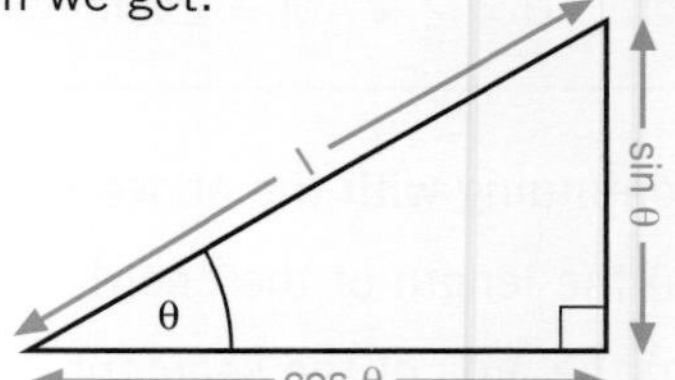

We can define the tangent function in terms of sin and cos.

$\tan\theta = \frac{\sin\theta}{\cos\theta}$

The graph of tan θ against θ looks like this:

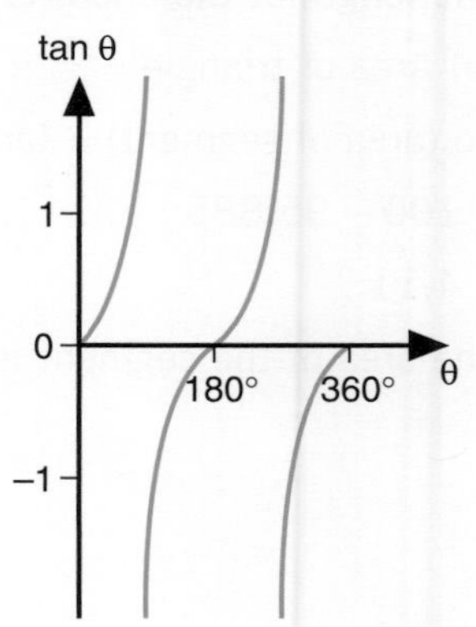

All these functions are **periodic** – that is, they repeat themselves. The sine and cosine functions repeat every 360° or 2π radians. The tangent function repeats every 180° or π radians.

The sine and cosine graphs are **continuous**. The tangent graph is **discontinuous** as it is not defined for angles θ where $\cos\theta = 0$. (You cannot divide by zero.)

As the angle moves through the four quadrants, the signs of the ratios change. This can be summarised by showing which ratios are positive in each quadrant:

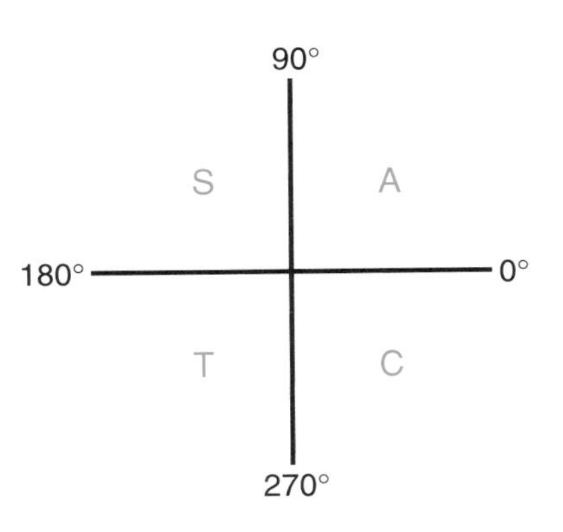

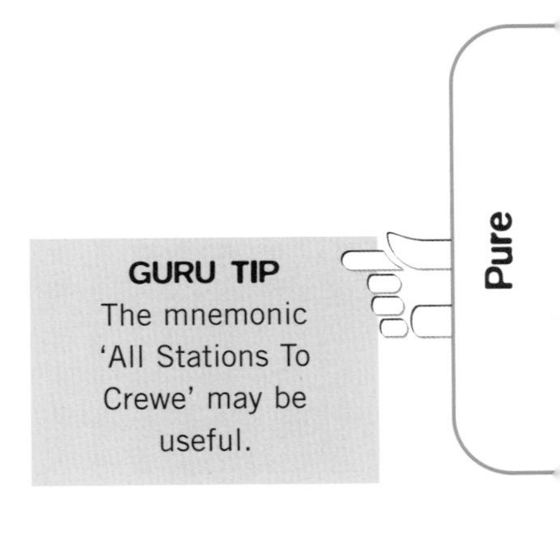

GURU TIP
The mnemonic 'All Stations To Crewe' may be useful.

Trigonometric functions of some common angles

Degrees	0	30	45	60	90
Radians	0	$\frac{\pi}{6}$	$\frac{\pi}{4}$	$\frac{\pi}{3}$	$\frac{\pi}{2}$
Sine	0	$\frac{1}{2}$	$\frac{1}{\sqrt{2}}$	$\frac{\sqrt{3}}{2}$	1
Cosine	1	$\frac{\sqrt{3}}{2}$	$\frac{1}{\sqrt{2}}$	$\frac{1}{2}$	0
Tangent	0	$\frac{1}{\sqrt{3}}$	1	$\sqrt{3}$	

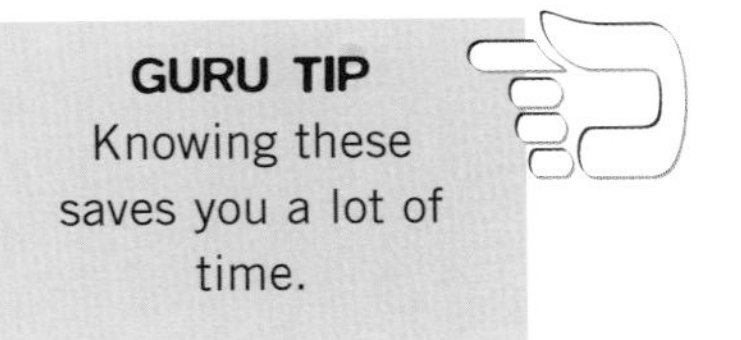

GURU TIP
Knowing these saves you a lot of time.

Trigonometry 2 – trigonometric equations

If you want to find an angle given the value of one of the trigonometric ratios, you need to solve a trigonometric equation. You will have to be able to work in degrees and radians.

EXAMPLE Find all values of θ between 0 and +360° such that $\sin\theta = \frac{1}{2}$.

SOLUTION First find the solution in the first quadrant using the inverse function:

$\theta = \sin^{-1}\frac{1}{2} = 30°$

Draw a sketch of the four quadrants. Show the angle in the first quadrant.

WEB TIP
There's more about Trignometric equations on the AS Guru™ website.

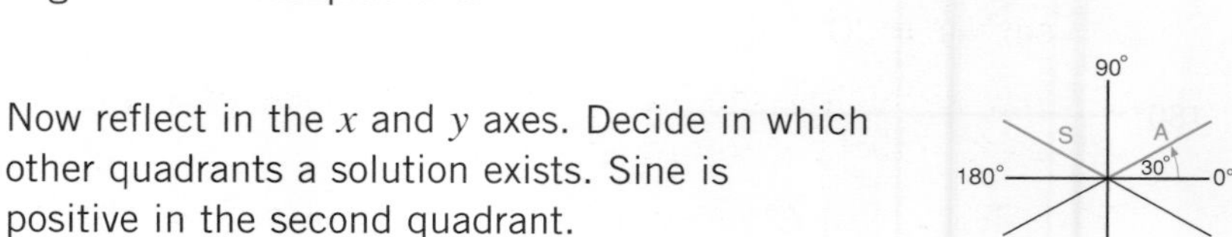

Now reflect in the x and y axes. Decide in which other quadrants a solution exists. Sine is positive in the second quadrant.

The angle in that quadrant, measured from 0°, is 180° – 30° = 150°

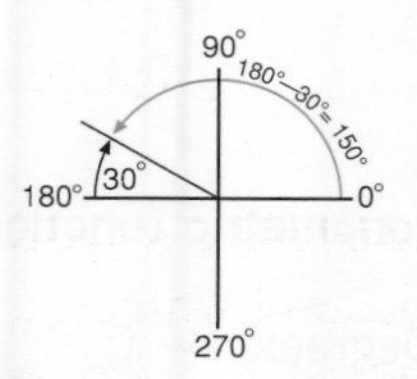

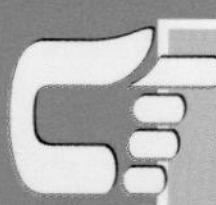

GURU TIP
If the question uses degrees, your answer should use degrees too. If the question uses radians, give your answer in radians.

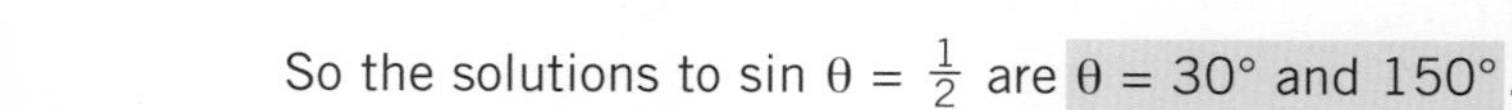

So the solutions to $\sin\theta = \frac{1}{2}$ are $\theta = 30°$ and $150°$.

What about negative sines, cosines or tangents?

EXAMPLE Find all values of θ between 0 and $+2\pi$ such that $\cos\theta = -\frac{1}{2}$.

SOLUTION Ignoring the sign, find the solution in the first quadrant.

$\theta = \cos^{-1}\frac{1}{2} = \frac{\pi}{3}$

Draw a sketch of the four quadrants. Show the angle in the first quadrant and reflect in the x and y axes.

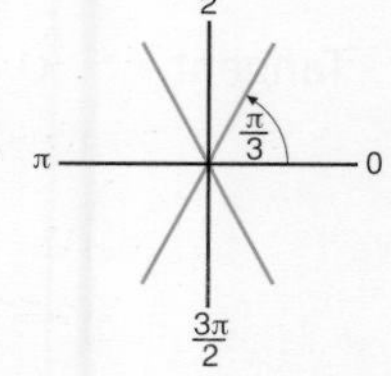

Now decide in which quadrants a solution exists. Cosine is negative in the second and third quadrants.

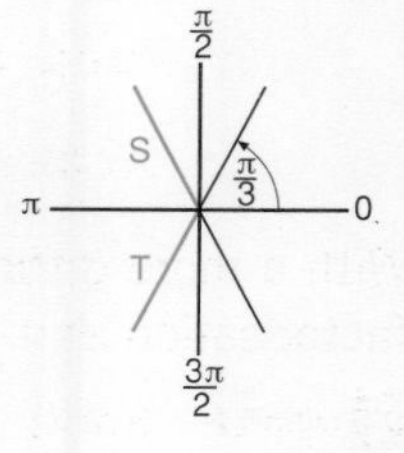

Find the angle in those quadrants measuring from 0°.

The solution in the second quadrant is $\pi - \frac{\pi}{3} = \frac{2\pi}{3}$, and the solution in the third quadrant is $\pi + \frac{\pi}{3} = \frac{4\pi}{3}$.

So the solutions to $\cos\theta = -\frac{1}{2}$ are $\theta = \frac{2\pi}{3}$ and $\frac{4\pi}{3}$.

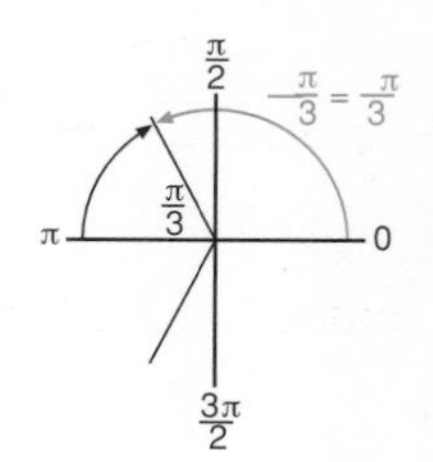

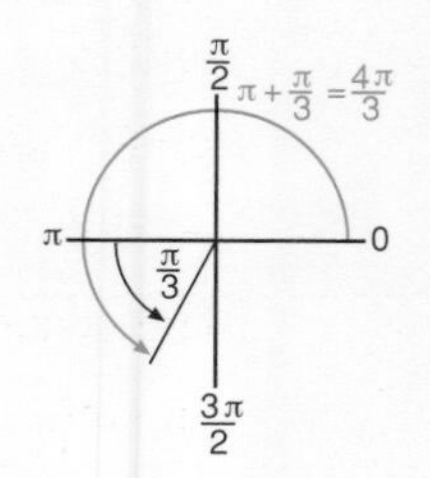

EXAMPLE Find all values of θ between $-\pi$ and $+\pi$ such that $\cos\theta = -\frac{1}{2}$.

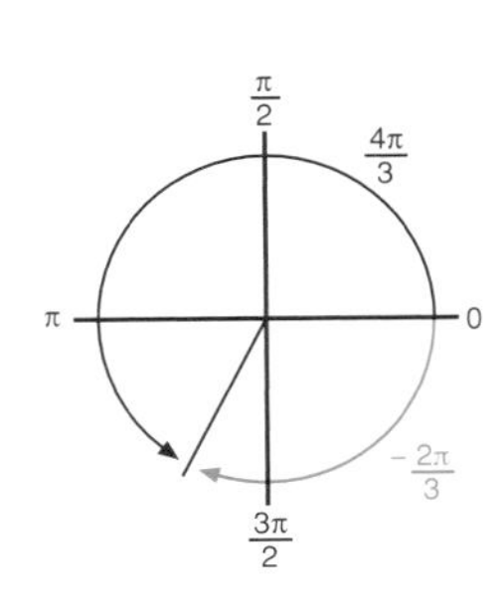

SOLUTION Proceed as above, but with the addition of a step to transform the solutions into the required range.

As the second solution is outside the range, subtract 2π.

The solutions are $\theta = \frac{2\pi}{3}$ and $-\frac{2\pi}{3}$.

EXAMPLE Solve $\sin x = 3\sin x - 1$ in the range $0 \le x \le 360°$.

SOLUTION 1. First solve for $\sin x$:

$\sin x = 3\sin x - 1 \Leftrightarrow 2\sin x = 1 \Leftrightarrow \sin x = \frac{1}{2}$.

2. Find the solution to this in the first quadrant:

$x = \sin^{-1}\frac{1}{2} = 30°$

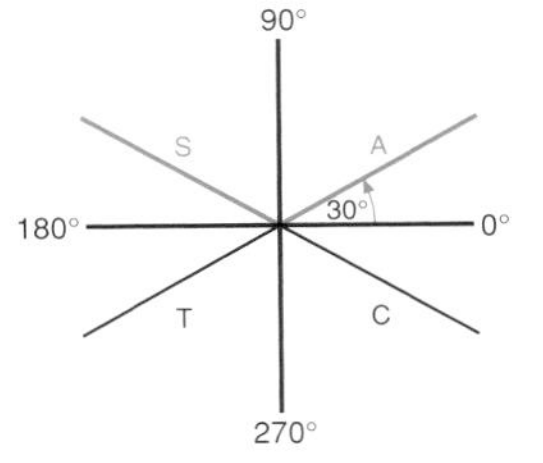

3. Draw a sketch of the four quadrants and show the angle in the first quadrant. Reflect in the x and y axes.

Now decide in which quadrants a solution exists. Sine is positive in the first and second quadrants.

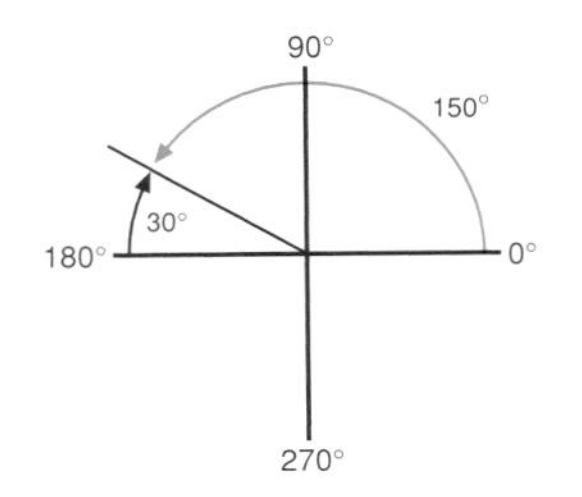

4. Find the angle in those quadrants measured from 0°.

5. The solutions are $x = 30°$ and $150°$.

Quadratic equations

If you start with a quadratic equation you will normally have two trigonometric equations to solve, and you'll end up with twice as many solutions.

EXAMPLE $\cos^2\theta = \frac{3}{4}$ $(0 \le \theta < 360°)$

SOLUTION 1. First solve for $\cos\theta$. Take square roots of both sides (and don't forget the negative root):

$\cos\theta = \pm\frac{\sqrt{3}}{2}$

2. Now $\cos^{-1}\frac{\sqrt{3}}{2} = 30°$.

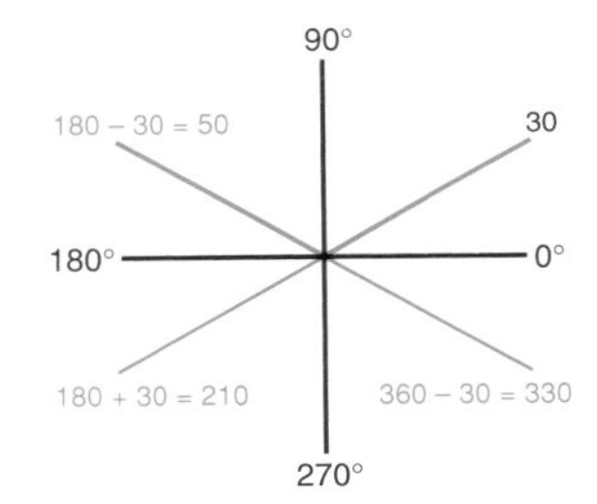

3. As the positive and negative values are allowed, there will be an answer in each quadrant:

4. If $\cos\theta = +\frac{\sqrt{3}}{2}$ then $\theta = 30°$ or $360° - 30° = 330°$.

If $\cos\theta = -\frac{\sqrt{3}}{2}$ then $\theta = 180° - 30° = 150°$ or $180° + 30° = 210°$.

5. So the solutions are $\theta = 30°, 150°, 210°, 330°$.

With a more difficult quadratic equation, you need to be able to solve by factorisation and by using the quadratic formula.

EXAMPLE Solve the equation $\sin^2\theta - \sin\theta - 2 = 0$ for $0 \le \theta < 360°$.

SOLUTION This equation in $\sin\theta$ factorises, so we can solve for $\sin\theta$, getting two solutions:

$\sin^2\theta - \sin\theta - 2 = 0 \Leftrightarrow (\sin\theta + 1)(\sin\theta - 2) = 0$

$\Leftrightarrow \sin\theta = -1$ or $\sin\theta = 2$

The second solution is impossible as $\sin\theta$ is always between ± 1.

So $\sin\theta = -1$.

Now $\sin^{-1}(1) = 90°$, and sine is negative in the third and fourth quadrants.

Therefore $\theta = 270°$.

GURU TIP
For quadratics in $\tan\theta$ there are no rejected solutions, as tan can take any value.

Differentiation 1 – first principles

On the website there is a lesson that starts with a plot of $y = x^2$ for positive values of x. You will be familiar with this curve.

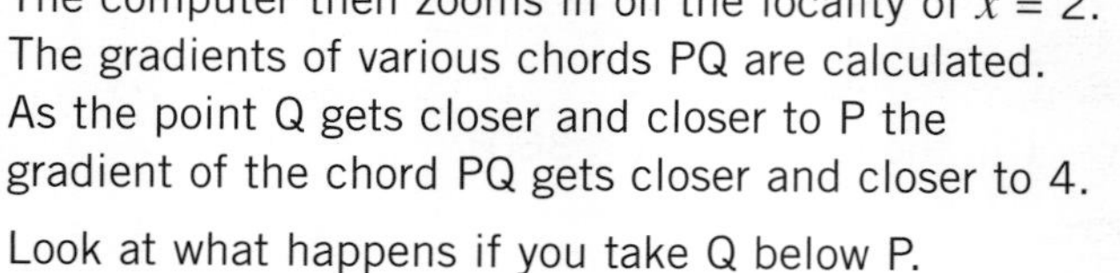

The computer then zooms in on the locality of $x = 2$. The gradients of various chords PQ are calculated. As the point Q gets closer and closer to P the gradient of the chord PQ gets closer and closer to 4.

Look at what happens if you take Q below P.

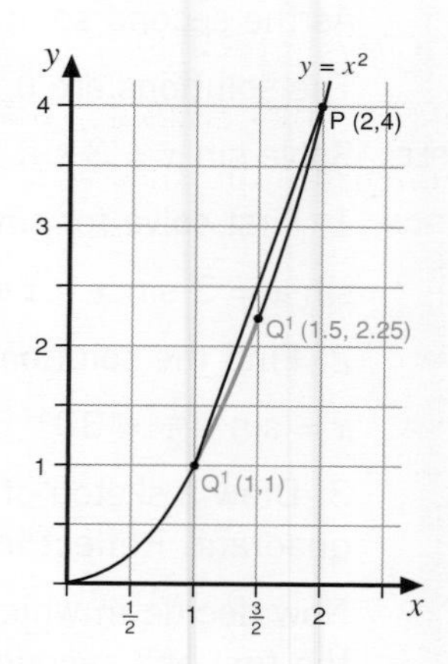

The gradients of the two chords shown are:

(1,1) to (2,4): gradient = $\frac{4-1}{2-1} = 3$

(1.5,2.25) to (2,4): gradient = $\frac{4-2.25}{2-1.5} = 3.5$

The second chord is a better approximation to the tangent. By zooming in on the curve we can obtain a still better approximation. Notice how the curve itself becomes more and more like a straight line as we zoom in on it.

GURU TIP
Work through these calculations to get a feel for the idea.

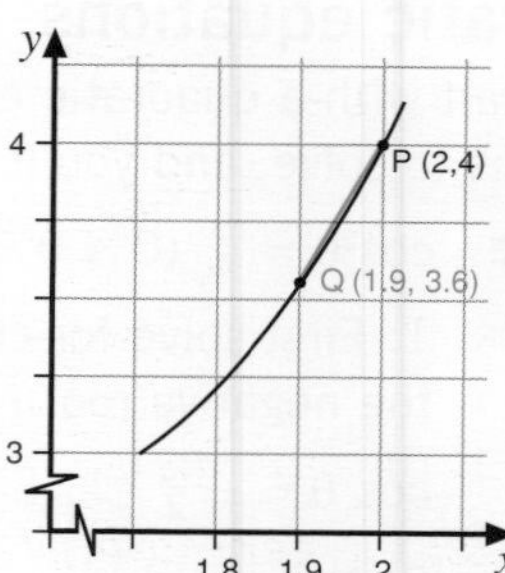
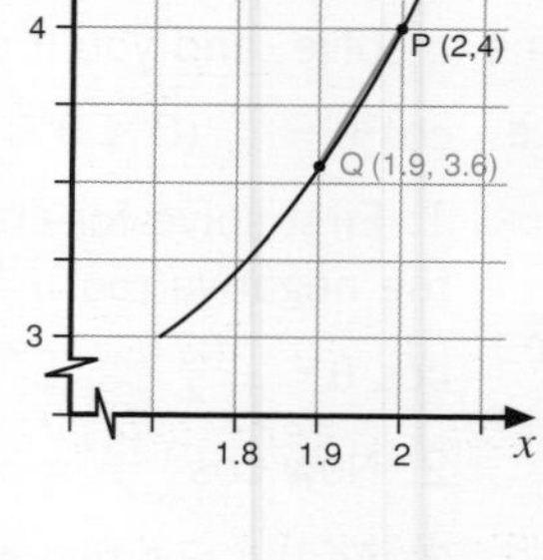

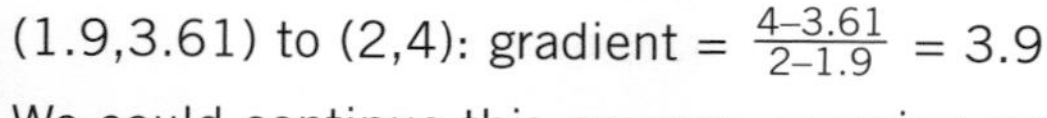

(1.9,3.61) to (2,4): gradient = $\frac{4-3.61}{2-1.9} = 3.9$

We could continue this process, zooming ever closer to the point (2,4). The gradient of the chord would get closer and closer to the gradient of the tangent, but would never quite reach it.

To formalise this sort of argument, we need the concept of a **limit**. We say:

As Q **tends to** P, the gradient of QP **tends to** 4.

WEB TIP
There's more about gradient on the AS Guru™ website.

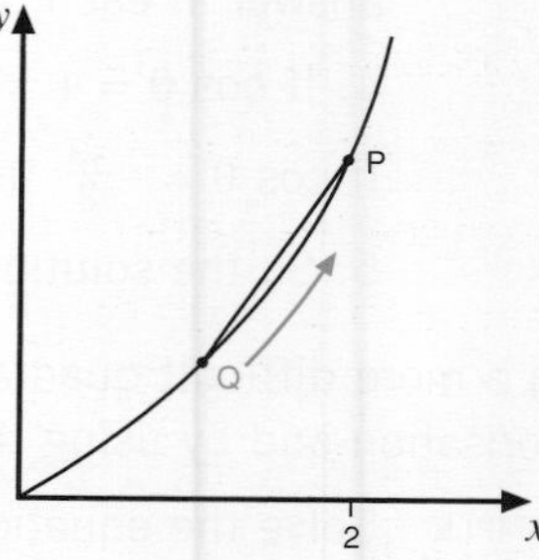

In other words, the **gradient of the curve** at $x = 2$ is 4.

So far we have just been calculating values. But to prove the above fact, we need to draw a general picture of a point moving Q closer and closer to P, this time from above.

Gradient of PQ $= \frac{(2+h)^2 - 4}{h}$

$= \frac{4 + 4h + h^2 - 4}{h}$

$= \frac{4h + h^2}{h}$

$= 4 + h$

So as h tends to 0, the gradient of PQ tends to 4.

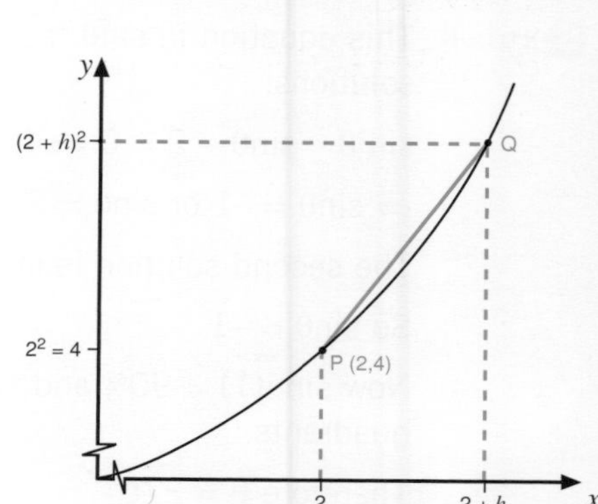

This process is called differentiating from first principles. You need to understand it, for two reasons. Firstly, it tells you *why* the rules work. Secondly, you could be asked to differentiate a simple function 'from first principles' – and you would need to show this type of working.

You now have a method to find the gradient at a given point on the curve $y = x^2$.

The function that gives the gradient at any value of x is called the **gradient function**.

Continuing with the curve $y = x^2$:

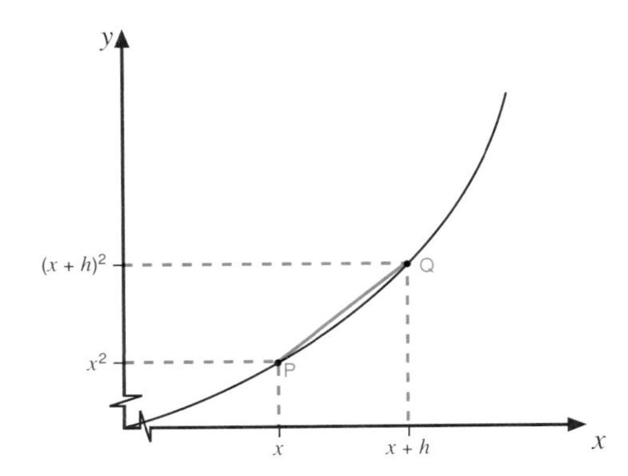

Let P be the point (x, x^2) and Q be the point $(x + h, (x + h)^2)$.

$$\text{Gradient of PQ} = \frac{(x + h)^2 - x^2}{(x + h) - x}$$
$$= \frac{x^2 + 2hx - x^2}{h}$$
$$= \frac{h(2x + h)}{h}$$
$$= 2x + h$$

As h tends to 0 (so Q tends to P), the gradient tends to $2x$.

We say that the **derivative** of x^2 with respect to x is $2x$.

Try repeating this for $y = x^3$. Let P be the point (x, x^3) and Q be the point $(x + h, (x + h)^3)$. You will need to use the binomial expansion for $(x + h)^3$.

You should find that the gradient of the tangent at P is $3x^2$.

For $y = x^4$ you would find that the gradient of the tangent is $4x^3$.

Can you see the pattern? If $y = x^n$ then the gradient of the tangent is nx^{n-1}.

In general, if $y = \mathrm{f}(x)$ then the gradient function is called the **derivative** of y **with respect to** x, and written $\frac{dy}{dx}$ or $\mathrm{f}'(x)$. Using this notation:

$$y = x^n \Rightarrow \frac{dy}{dx} = nx^{n-1}$$

Alternatively:

$$\mathrm{f}(x) = x^n \Rightarrow \mathrm{f}'(x) = nx^{n-1}$$

Pure

Differentiation 2 – the rules

GURU TIP
Learn these rules.

$y = x^n \Rightarrow \frac{dy}{dx} = nx^{n-1}$

$y = kx^n \Rightarrow \frac{dy}{dx} = nkx^{n-1}$

where k is any constant

$y = c \Rightarrow \frac{dy}{dx} = 0$

where c is any constant

The last equation makes sense because the line $y = c$ is horizontal and so has gradient 0.

Using the rules

GURU TIP
'Multiply the coefficient by the power and take one from the power.'

EXAMPLE Find the derivatives of the following functions:

(a) $y = 3x^5$

(b) $f(x) = 6x^4$

SOLUTION (a) $\frac{dy}{dx} = 5 \times 3x^{5-1} = 15x^4$

(b) $f'(x) = 4 \times 6x^{4-1} = 24x^3$

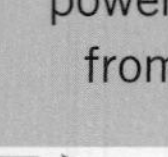

WEB TIP
There's more about differentiating polynomials on the AS Guru™ website.

To differentiate polynomials, you just differentiate term by term.

In any question, 'differentiate...' means 'find the derivative of...'.

EXAMPLE (a) Differentiate with respect to x: $y = 6x^2 + 2x + 6$.

(b) Differentiate with respect to x: $f(x) = 5x^4 - 6x^3 - x^2 + 10x - 5$.

(c) Find the gradient of the curve $y = x^3 - x^2 - 2x$ at the point (2,0).

SOLUTION (a) $\frac{dy}{dx} = 2 \times 6x^{2-1} + 1 \times 2x^{1-1} = 12x + 2$.

(b) $f'(x) = 4 \times 5x^{4-1} - 3 \times 6x^{3-1} - 2x^{2-1} + 10 = 20x^3 - 18x^2 - 2x + 10$.

(c) $\frac{dy}{dx} = 3x^{3-1} - 2x^{2-1} - 2 = 3x^2 - 2x - 2$.

Substitute $x = 2$: $\frac{dy}{dx} = 3 \times 2^2 - 2 \times 2 - 2 = 6$.

GURU TIP
If $y = ax$ then $\frac{dy}{dx} = a$.

Having found the gradient of the curve at the point (2,0), you can go on to find the equation of the tangent at this point.

The equation of a tangent to a curve

The equation of the tangent to the curve $y = f(x)$ at (x_1, y_1) is

$y - y_1 = m(x - x_1)$

where $m = f(x_1)$.

EXAMPLE Find the equation of the tangent to the curve $y = x^3 - x^2 - 2x$ at the point (2,0).

SOLUTION Use the above formula with $x_1 = 2$, $y_1 = 0$. We have already found that $m = f'(2) = 2$ (see above). So the equation is

$y - 0 = 2(x - 2)$

or $y = 2x - 4$.

You may also need to find the equation of the normal.

> **GURU TIP**
> This is what you did on page 13 with a straight line.

The equation of a normal to a curve

The key to finding the equation of the normal is that the normal is perpendicular to the tangent.

Remember that the product of the gradients of two perpendicular lines is –1. So if the gradient of the tangent is m_1 and the gradient of the normal is m_2:

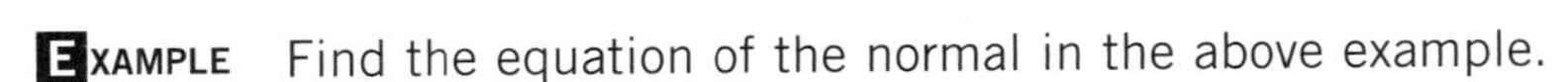

> The equation of the normal to the curve $y = f(x)$ at (x_1, y_1) is
> $y - y_1 = m_2(x - x_1)$
> where $m_2 = -\frac{1}{m_1} = -\frac{1}{f'(x_1)}$.

Pure

EXAMPLE Find the equation of the normal in the above example.

SOLUTION The gradient of the tangent $m_1 = 2$, so the gradient of the normal $m_2 = -\frac{1}{m_1} = -\frac{1}{2}$.

The equation of the normal is

$y - 0 = -\frac{1}{2}(x - 2)$

$\Leftrightarrow y = -\frac{1}{2}x + 1$

EXAMPLE Find the equation of the tangent and the normal to the curve $y = 2x^3 - 4x$ at the point on the curve where $x = 1$.

SOLUTION First, to find the gradient of the tangent, differentiate the function $y = f(x) = 2x^3 - 4x$:

$f'(x) = 6x^2 - 4 \Rightarrow f'(1) = 2$.

Now $x = 1 \Rightarrow y = -2$. So the equation of the tangent is

$y - (-2) = 2(x - 1)$

$\Leftrightarrow y = 2x - 4$

Now the gradient of the normal is $m_2 = -\frac{1}{m_1} = -\frac{1}{2}$. So the equation of the normal is

$y - (-2) = -\frac{1}{2}(x - 1)$

$\Leftrightarrow y = -\frac{1}{2}x - \frac{3}{2}$

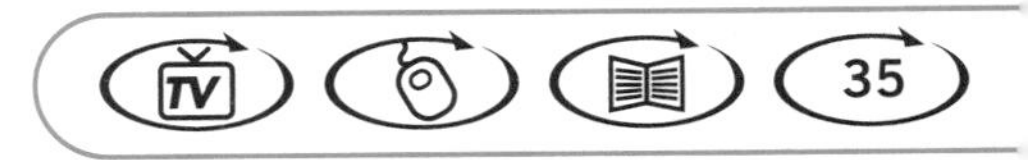

Differentiation 3 – maxima and minima

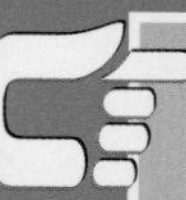

GURU TIP
At any maximum or minimum point, the gradient is 0. That is, the derivative of the function at that point is 0.

How do we find the minimum point of the graph $y = x^2 - 10x + 21$?

The derivative of the function, $\frac{dy}{dx} = 2x - 10$, gives us the gradient at any point.

For example, the gradient at $x = 0$ is –10, and the gradient at $x = 6$ is 2.

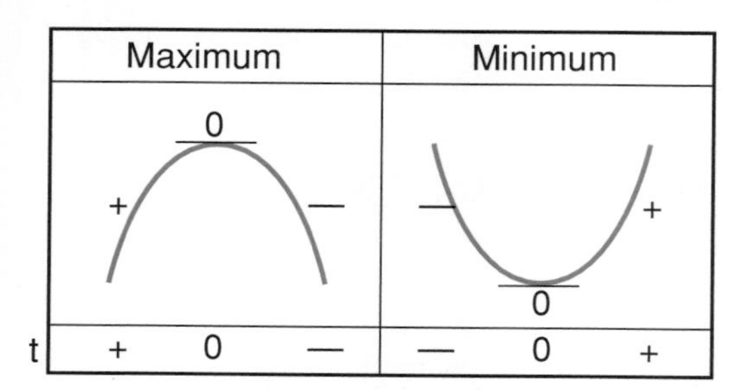

A point at which the gradient is 0 is called a **stationary point**.

So $y = x^2 - 10x + 21$ has a maximum or minimum value at the point where $\frac{dy}{dx} = 2x - 10 = 0 \Leftrightarrow x = 5$.

As we have already calculated $f'(0) = -10$ and $f'(6) = 2$, from the picture above we can see that this point must be a minimum.

GURU TIP
For a quadratic function, you can also find the stationary point by completing the square.

EXAMPLE Find the coordinates of the stationary points of $y = x^3 - 6x^2 + 9x + 2$, stating whether they are maximum or minimum points.

SOLUTION 1. First differentiate the function:

$\frac{dy}{dx} = 3x^2 - 12x + 9$

2. To find the values of x at the stationary points, solve $\frac{dy}{dx} = 0$:

$3x^2 - 12x + 9 = 0 \Leftrightarrow (3x - 3)(x - 3) = 0 \Leftrightarrow 3(x - 1)(x - 3) = 0$
$\Leftrightarrow x = 1 \text{ or } 3.$

3. Next find the corresponding y values:

$x = 1 \Rightarrow y = 1^3 - 6\times1^2 + 9\times1 + 2 = 6$

$x = 3 \Rightarrow y = 3^3 - 6\times3^2 + 9\times3 + 2 = 2$

4. Finally decide whether each point is a maximum or minimum. Look at the gradients of points either side of the stationary points:

$x = 0 \Rightarrow \frac{dy}{dx} = 3\times0^2 - 12\times0 + 9 = 9 > 0$

$x = 2 \Rightarrow \frac{dy}{dx} = 3\times2^2 - 12\times2 + 9 = -3 < 0$

$x = 4 \Rightarrow \frac{dy}{dx} = 3\times4^2 - 12\times4 + 9 = 9 > 0$

x	0	1	2
$\frac{dy}{dx}$	+	0	–
	/	—	\

The gradient goes '+ 0 –', so the point (1,6) is a maximum.

x	2	3	4
$\frac{dy}{dx}$	–	0	+
	\	—	/

The gradient goes '– 0 +', so the point (3,2) is a minimum.

WEB TIP
There's more about Stationary points on the AS Guru™ website.

Problems involving maxima and minima

EXAMPLE This cuboid has a surface area of 300cm²:

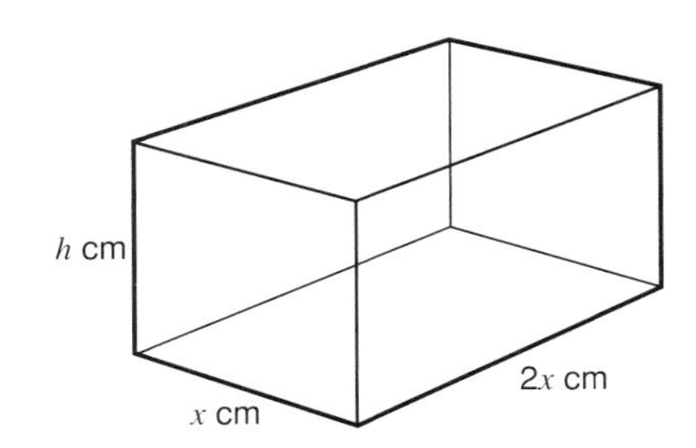

(a) Express h in terms of x.

(b) Show that the volume V cm³ of the cuboid is given by

$V = 100x - \frac{4x^2}{3}$

(c) Find the value of x which gives the maximum volume. Explain why this is a maximum.

SOLUTION (a) The surface area is $2(x \times 2x + 2x \times h + h \times x) = 4x^2 + 6hx$.

$4x^2 + 6hx = 300$

$\Rightarrow h = \frac{300 - 4x^2}{6x}$

(b) $V = x \times 2x \times h$

$= 2x^2 \; \frac{300 - 4x^2}{6x} = \frac{x}{3}(300 - 4x^2) = 100x - \frac{4x^2}{3}$ as required

(c) For a maximum or minimum,

$\frac{dV}{dx} = 0 \Leftrightarrow 100 - 4x^2 = 0$

for a max or min, $\frac{dV}{dx} = 0$

$\Rightarrow 100 - 4x^2 = 0$

$\Leftrightarrow 4x^2 = 100$

$\Leftrightarrow x^2 = 25$

$\Leftrightarrow x = 5$ (as $x > 0$)

To test for maximum or minimum:

x	4	5	6
$\frac{dy}{dx}$	+	0	–
	/	—	\

The solution gives the maximum volume.

GURU TIP
Practise lots of questions like this. Examiners love them!

WEB TIP
There's more about Maxima and minima on the AS Guru™ website.

Pure

Integration 1 – principles

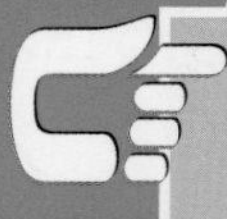

GURU TIP
Integration is the reverse of differentiation.

The opposite of differentiation

The website has an activity in which you have to match a derivative with the appropriate function. For example, given $f'(x) = 2x$ you match it with $f(x) = x^2$. You are undoing the process of differentiating the function. This reverse process – the opposite of differentiation – is called **integration**.

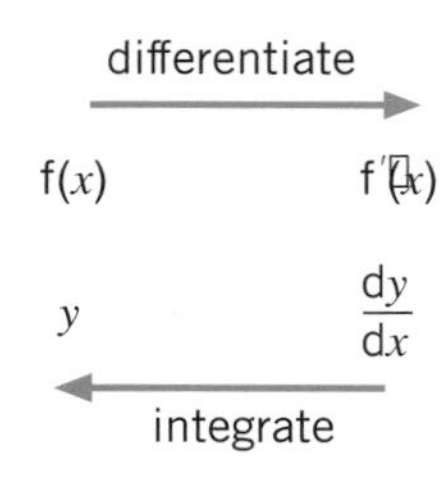

The indefinite integral

More than one function can have the same derivative. For example, $\frac{dy}{dx} = 2$ for both $y = 2x + 4$ and $y = 2x - 2$. In fact, any function of the form $y = 2x + c$ would satisfy $\frac{dy}{dx} = 2$.

WEB TIP
There's more about the Definite integral on the AS Guru™ website.

The number c is called the **constant of integration**.

Another way of expressing

$\frac{dy}{dx} = 2 \Rightarrow y = 2x + c$

is

$\int 2\,dx = 2x + c$

We say that the (**indefinite**) **integral** of 2 with respect to x is $2x + c$.

Differential equations

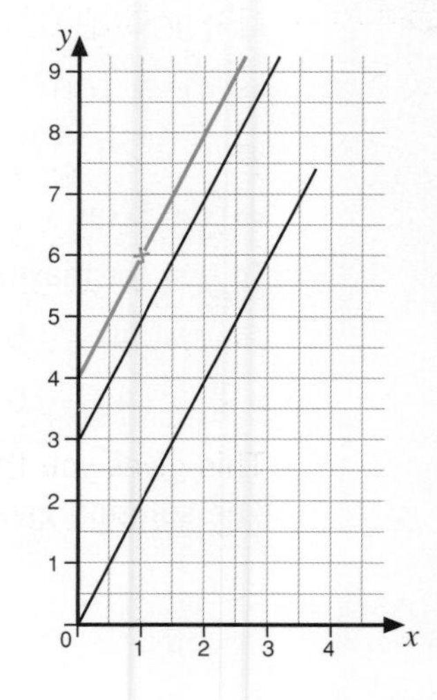

A differential equation is an equation that includes a derivative, such as $\frac{dy}{dx}$. $\frac{dy}{dx} = 2$ is an example of a differential equation.

The indefinite integral $y = 2x + c$ is called the **general solution**. It may be drawn as a family of parallel straight lines.

You may be given a condition such as 'the value of y is 6 at $x = 1$'. This is called a **boundary condition**. You can use it to evaluate c. In this case it would tell you that $c = 4$, giving the **particular solution** $y = 2x + 4$.

Differential equations are very important in physics, economics, biology and countless other subjects.

Integration rules

You need to be able to integrate constants and powers of x.

$$\frac{dy}{dx} = a \Rightarrow y = ax + c$$

$$\frac{dy}{dx} = x^n \Rightarrow y = \frac{x^{n+1}}{n+1} + c \quad (n \neq -1)$$

For some awarding bodies you only need to integrate integer powers; for others you need to integrate fractional powers. The rule works for both. It does not, however, work for $\frac{dy}{dx} = \frac{1}{x} = x^{-1}$, so you will not be asked to do this in the first pure maths module.

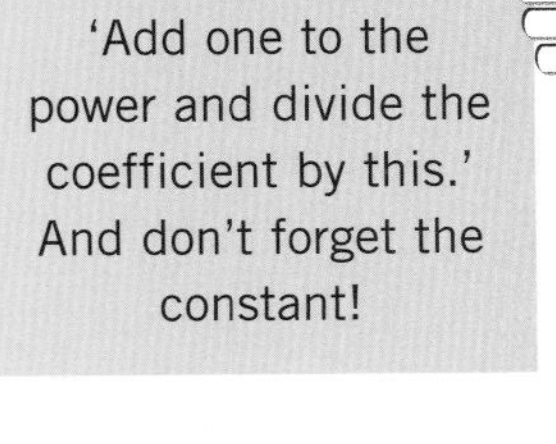

GURU TIP
'Add one to the power and divide the coefficient by this.' And don't forget the constant!

Examples

If $\frac{dy}{dx} = 12x$ then $y = 12\frac{x^{1+1}}{1+1} + c = 6x^2 + c$

If $\frac{dy}{dx} = -3$ then $y = -3x + c$

If $\frac{dy}{dx} = x^{-2}$ then $y = \frac{x^{-2+1}}{-2+1} + c = -x^{-1} + c$

If $\frac{dy}{dx} = x^{\frac{3}{2}}$ then

$$y = \frac{x^{\frac{3}{2}+1}}{\frac{3}{2}+1} + c = \frac{x^{\frac{5}{2}}}{\frac{5}{2}} + c = \frac{2x^{\frac{5}{2}}}{5} + c$$

Pure

Integrating polynomials

To integrate a polynomial you just apply the above rules term by term.

WEB TIP
There's more about Integrating polynomials on the AS Guru™ website.

Example

If $\frac{dy}{dx} = 3x^2 - 4x + 2$ then $y = \int 3x^2\, dx - \int 4x\, dx + \int 2\, dx$

$y = \frac{3x^{2+1}}{2+1} - \frac{4x^{1+1}}{1+1} + 2x + c$

$y = x^3 - 2x^2 + 2x + c$

Finding the equation of a curve given the gradient function and a point

EXAMPLE Find the equation of the curve that passes through (1,4) and has gradient function $\frac{dy}{dx} = 2x - 4$.

SOLUTION First integrate the gradient function to find the general solution of the differential equation.

$\frac{dy}{dx} = 2x - 4$

$\Rightarrow y = x^2 - 4x + c$

This gives you a family of curves.

Next, substitute in the values for x and y given by the point (the boundary condition) to find c.

$4 = 1^2 - 4 \times 1 + c \Rightarrow c = 7$

This gives you the particular solution required:

$y = x^2 - 4x + 7$

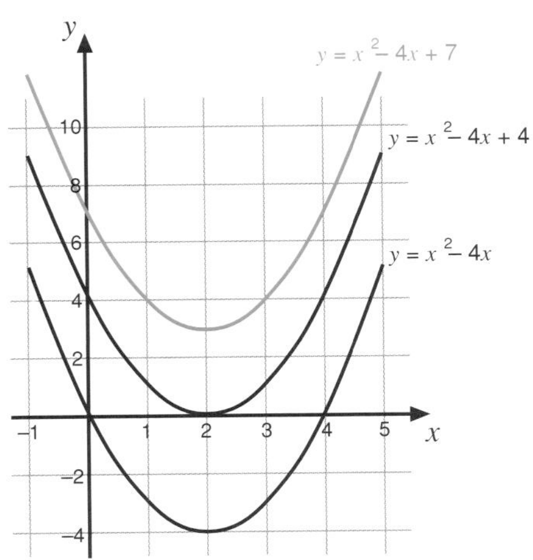

Integration 2 – area under a curve

The definite integral

You may be asked to evaluate integrals between limits. A constant of integration is not required in this case as it would just be cancelled out. The answer is a numerical value and is called a **definite integral**.

The upper limit of a definite integral is written above the symbol $\int$ and the lower limit below it.

EXAMPLE Evaluate $\int_1^4 x^2 dx$

SOLUTION Integrate the function, and subtract the value at the lower limit from the value at the upper limit:

$$\int_1^4 x^2 dx = \left[\frac{x^3}{3}\right]_1^4 = \frac{4^3}{3} - \frac{1^3}{3} = \frac{64}{3} = 21.$$

WEB TIP
There's more about finding the area under a graph on the AS Guru™ website.

Finding the area under a curve

The integral of a function can be used to find the area under a curve.

In the above example, we have found the area under the curve $y = x^2$ between $x = 1$ and $x = 4$.

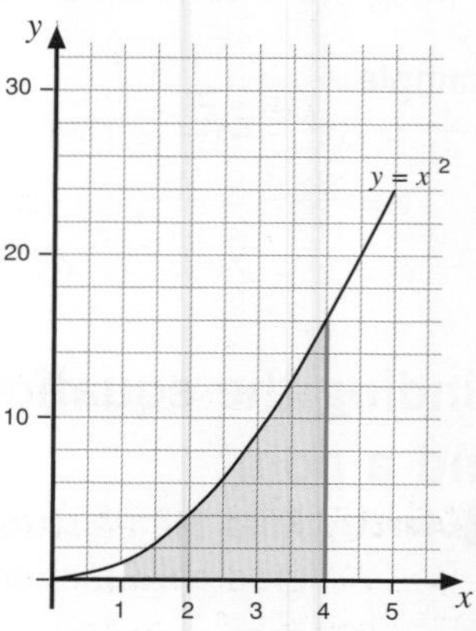

The area between the curve $y = f(x)$ and the x axis, between $x = a$ and $x = b$, is

$$\int_a^b f(x)dx = \left[F(x)\right]_a^b = F(b) - F(a)$$

where $F(x) = \int f(x)\, dx$

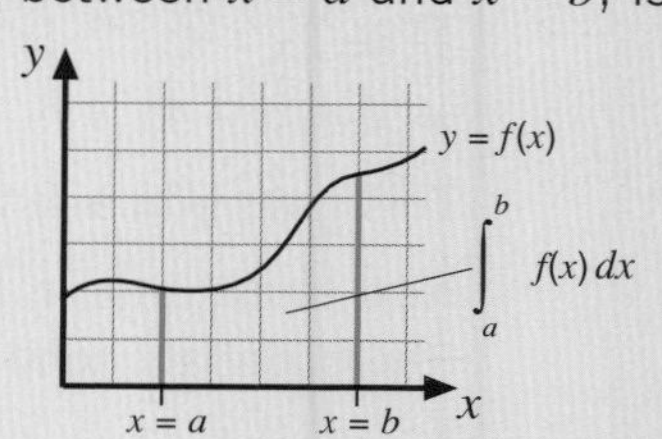

Area under the x axis

EXAMPLE Find the area bounded by the curve $y = x^2 - 5x$ between $x = 2$ and $x = 3$.

SOLUTION Area $= \int_2^3 (x^2 - 5x)dx = \left[\frac{x^3}{3} - \frac{5x^2}{2}\right]_2^3$

$= \left(\frac{3^3}{3} - \frac{5\times3^2}{2}\right) - \left(\frac{2^3}{3} - \frac{5\times2^2}{2}\right)$

$= \frac{27-8}{3} - \frac{45-20}{2}$

$= -\frac{37}{6}$

The result is negative because the curve is under the axis.

The area under the x axis is $\frac{37}{6}$.

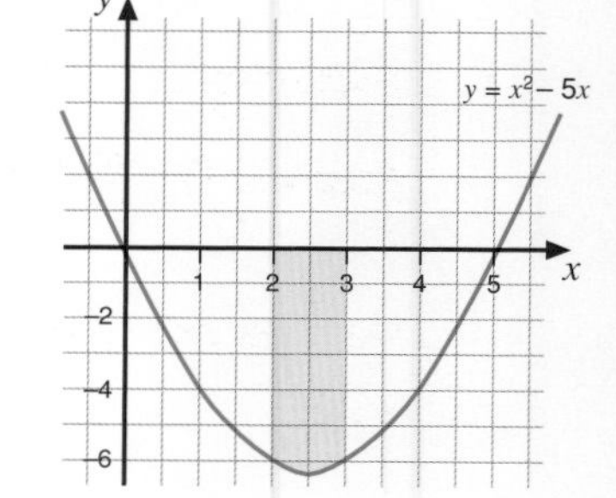

GURU TIP
Always sketch the function before evaluating the integral.

More complicated areas

EXAMPLE Find the area between the curve $y = x^2 - 5x$ and the x axis between $x = -2$ and $x = 3$.

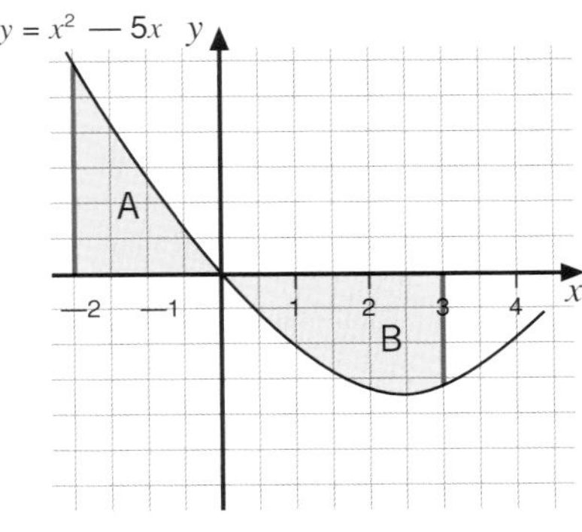

SOLUTION Sketch the graph.

Notice that the curve is:

- above the x axis between $x = -2$ and $x = 0$
- below the x axis between $x = 0$ and $x = 3$

To find the area between the graph and the x axis you must calculate the two areas separately, and then add them together, ignoring the minus signs.

$$\text{Area A} = \int_{-2}^{0} (x^2 - 5x)dx = \left[\frac{x^3}{3} - \frac{5x^2}{2}\right]_{-2}^{0}$$

$$= 0 - \left(\frac{(-2)^3}{3} - \frac{5\times(-2)^2}{2}\right)$$

$$= \frac{38}{3}$$

$$\text{Area B} = \int_{0}^{3} (x^2 - 5x)dx = \left[\frac{x^3}{3} - \frac{5x^2}{2}\right]_{0}^{3}$$

$$= \left(\frac{3^3}{3} - \frac{5\times3^2}{2}\right) - 0$$

$$= -\frac{27}{2}$$

$$\text{Total area} = \frac{38}{3} + \frac{27}{2} = \frac{157}{6}.$$

Finding the area between a curve and the y axis

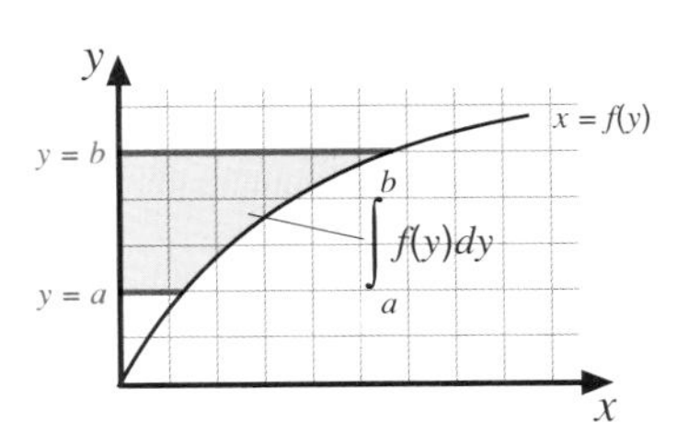

To find the area between a curve and the y axis, you just rearrange the function so that you have x expressed as a function of y, and apply the same method as above, interchanging x and y.

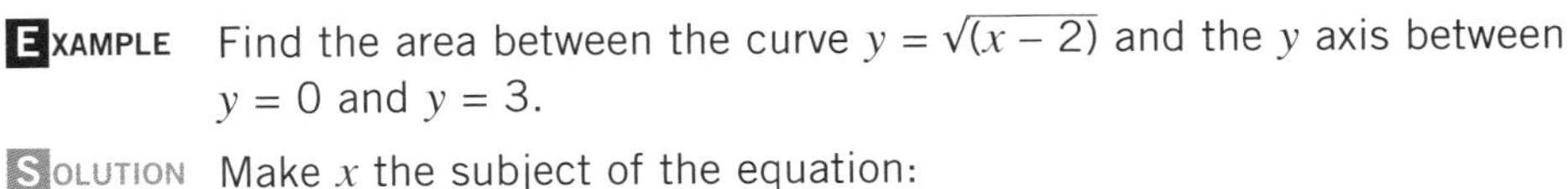

EXAMPLE Find the area between the curve $y = \sqrt{(x - 2)}$ and the y axis between $y = 0$ and $y = 3$.

SOLUTION Make x the subject of the equation:

$y = \sqrt{(x - 2)} \Rightarrow y^2 = x - 2$

$\Rightarrow x = y^2 + 2$

Sketch the graph:

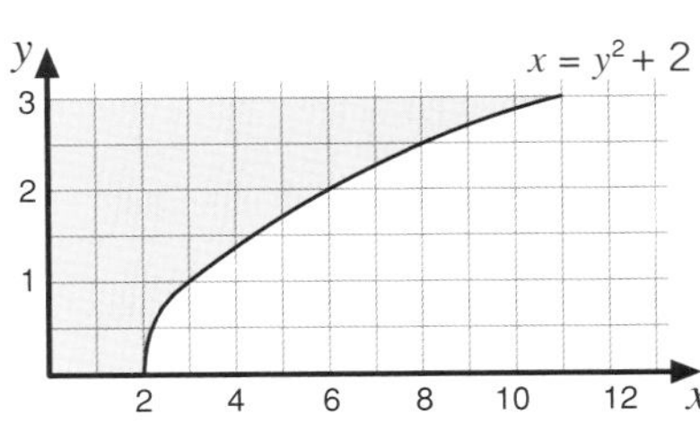

Integrate:

$$\text{Area} = \int_{0}^{3} (y^2+2)dy = \left[\frac{y^3}{3} + 2y\right]_{0}^{3}$$

$$= \left(\frac{3^3}{3} + 2\times3\right) - 0$$

$$= 15.$$

Integration 3 – further techniques

Finding the area between two curves

EXAMPLE Find the area enclosed by the curve $y = x^2 - 4x + 4$ and the line $y = x + 4$.

> If $f(x) \geq g(x)$ then the area between the curves $y = f(x)$ and $y = g(x)$ between $x = a$ and $x = b$ is
>
> $$\int_a^b f(x)dx - \int_a^b g(x)dx = \int_a^b f(x) - g(x)dx$$

SOLUTION First sketch the graphs and find the x values of the points where they intersect.

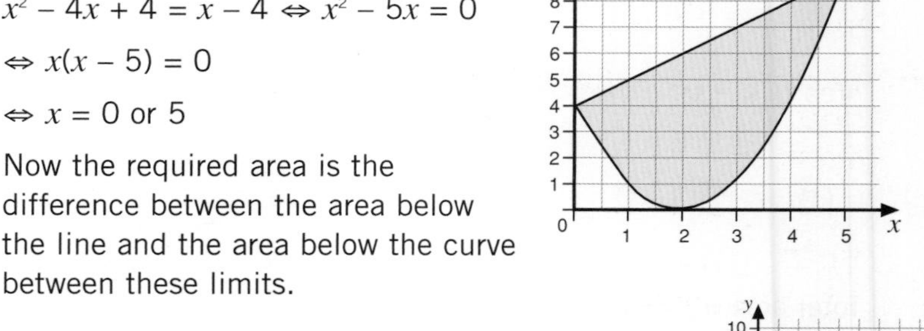

$x^2 - 4x + 4 = x - 4 \Leftrightarrow x^2 - 5x = 0$

$\Leftrightarrow x(x - 5) = 0$

$\Leftrightarrow x = 0$ or 5

Now the required area is the difference between the area below the line and the area below the curve between these limits.

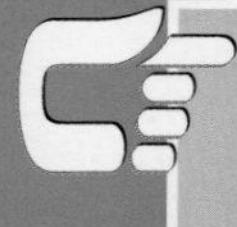

GURU TIP
When calculating the area below a straight line you can always calculate the area of the trapezium instead of integrating.

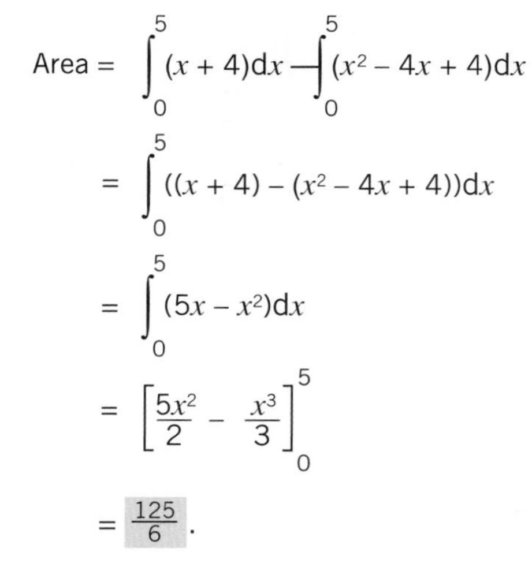

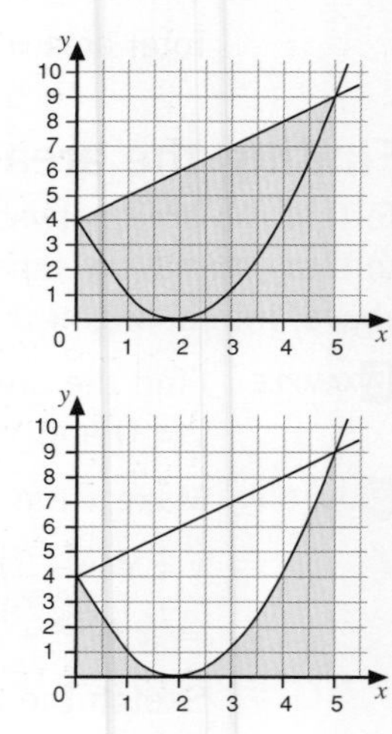

$$\text{Area} = \int_0^5 (x + 4)dx - \int_0^5 (x^2 - 4x + 4)dx$$

$$= \int_0^5 ((x + 4) - (x^2 - 4x + 4))dx$$

$$= \int_0^5 (5x - x^2)dx$$

$$= \left[\frac{5x^2}{2} - \frac{x^3}{3}\right]_0^5$$

$$= \frac{125}{6}.$$

EXAMPLE Find the area between the curve $x = y^2 + 1$ and the line $2y - x = 0$ for $0 \leq x \leq 2$, $0 \leq y \leq 1$.

SOLUTION 1. First sketch the graphs and find the points where they intersect.

$2y - x = 0 \Leftrightarrow x = 2y$

$2y = y^2 + 1 \Leftrightarrow y^2 - 2y + 1 = 0$

$\Leftrightarrow (y - 1)^2 = 0$

$\Leftrightarrow y = 1$, so $x = 2$.

Note the repeated root (the line is tangent to the curve at $x = 2$).

2. Now the required area is A – B.

$$\text{Area A} = \int_0^1 (y^2 + 1)dy$$

$$= \left[\frac{y^3}{3} + y\right]_0^1$$

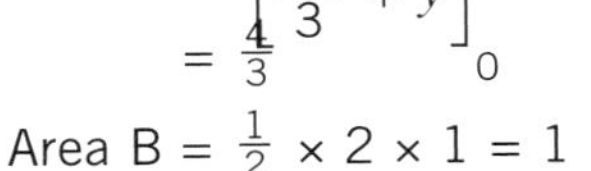

$$= \frac{4}{3}$$

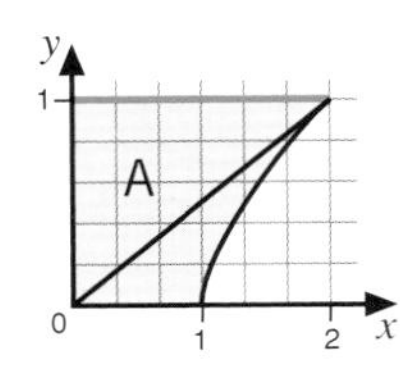

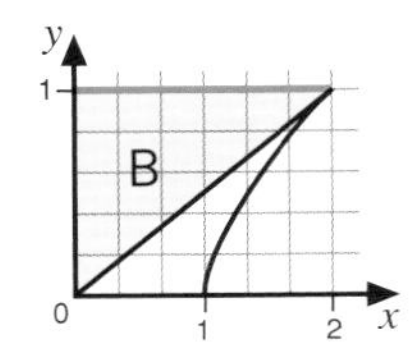

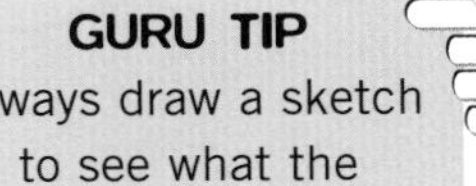

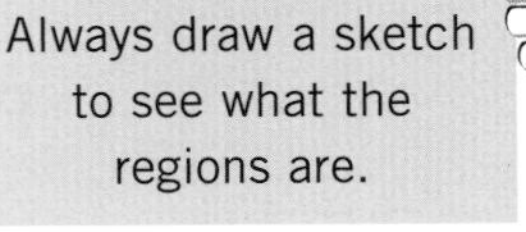

GURU TIP
Always draw a sketch to see what the regions are.

Area B $= \frac{1}{2} \times 2 \times 1 = 1$

So the required area is $\frac{4}{3} - 1 = \frac{1}{3}$

Pure

Numerical integration – the trapezium rule

Not all functions can be integrated using the rules we have. How do you calculate the area under a curve if you can't integrate the function?

One way is to divide the area into strips, and add up the areas of the trapezia to find an approximation to the area under the graph.

EXAMPLE Find the approximate area between the x axis and the curve $y = \sqrt{(25 - x^2)}$ between $x = 0$ and $x = 5$.

SOLUTION Sketch the curve. Divide it into strips and find the heights of the strips.

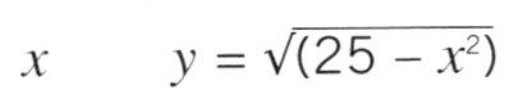

x	$y = \sqrt{(25 - x^2)}$
0	5.0000
1	4.8990
2	4.5826
3	4.0000
4	3.0000
5	0.0000

Find the areas of the trapezia:

A $= \frac{1}{2}(5.0000+4.8990) =$ 4.9495

B $= \frac{1}{2}(4.8990+4.5826) = 4.7408$

C $= \frac{1}{2}(4.5826+4.0000) = 4.2913$

D $= \frac{1}{2}(4.0000+3.0000) = 3.5000$

E $= \frac{1}{2}(3.0000+0.0000) = 1.5000$

Total = 18.9816.

As the area is that of a quarter circle, you could have used the formula for the area of a circle. If you do so you will see that the above estimate is fairly close. As you would expect from the diagram it is an underestimate.

In general, as you increase the number of strips, your estimate gets closer and closer to the true value. How quickly it converges depends on the shape of your curve.

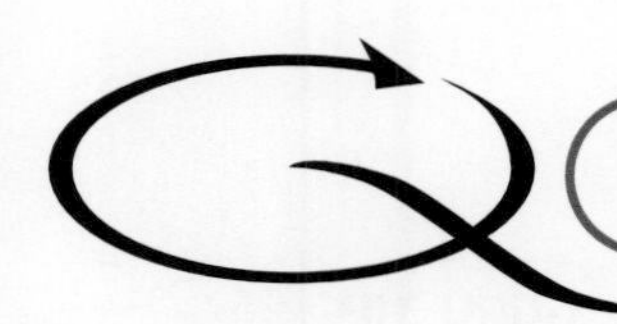

Practice questions

Proof

1 (a) Work out the value of $n^2 - n + 41$ for $n = 1, 2,..., 10$. Do you always get a prime number?
(b) Investigate whether $n^2 - n + 41$ is always prime for postive integers n.

2 Is the statement $2^n \geq n^2$ true for all whole-number values of n? If so, prove it. If not, modify the statement so that it is always true and prove it.

3 Prove that the product of four consecutive integers is always divisible by 24.

4 Prove that $\sin^2\theta + \cos^2\theta = 1$ for $0 \leq \theta \leq 90°$.

5 Prove that $n^5 - n^3$ is divisible by 24 for all integer n.

Indices and surds

6 Simplify:

(a) $x^3 \times x^4$ (c) $(2x^3)^7$ (e) $(x^{-7})^{-4}$

(b) $x^7 \div x^4$ (d) $x^4 \times \frac{1}{x^7}$ (f) $(x^4)^0$

7 Simplify:

(a) $8^{\frac{2}{3}}$ (c) $(49x^2)^{\frac{1}{2}}$ (e) $16^{-\frac{3}{4}}$

(b) $36^{-\frac{1}{2}}$ (d) $(3x^3)^3$ (f) $32^{0.8}$

8 Simplify:

(a) $\sqrt{5} \times \sqrt{5}$ (c) $\sqrt{2} \times \sqrt{32}$ (e) $\sqrt{5} + \sqrt{125}$

(b) $(\sqrt{5})^4$ (d) $\sqrt{27} \div 3\sqrt{3}$ (f) $(1 + \sqrt{3})^2$

9 Rationalise:

(a) $\frac{3}{\sqrt{5}}$ (c) $\frac{\sqrt{7}}{3\sqrt{2}}$

(b) $\frac{\sqrt{5}}{\sqrt{3}}$ (d) $\frac{1+\sqrt{2}}{\sqrt{2}}$

The equation of a straight line

10 Which of the following equations describe straight lines?

(a) $3x + y = 6$ (d) $y - 3\sqrt{x} = 4$

(b) $3xy = 5$ (e) $y + 3x = 0$

(c) $3x^2 - y = 0$ (f) $\frac{y}{x} - 2 = 0$

11 Find the gradients of the straight lines joining the following pairs of points:

(a) (1,3) and (2,8) (b) (–3,2) and (2,–3) (c) (2,–4) and (–3,6)

12 Find the coordinates of the midpoints of the lines joining the following points:

(a) (1,3) and (2,8) (b) (–3,2) and (2,–3) (c) (2,–4) and (–3,6)

13 Find the gradients of the straight lines with the following equations:

(a) $y = 3x - 7$ (b) $y = 6 - 2x$ (c) $5y + x = 7$

Mechanics – introduction

In mechanics you will be using mechanical principles to set up mathematical equations modelling a situation. The setting up of these equations attracts the majority of the examiner's marks. If they are set up incorrectly, the resulting calculations will be wrong.

Clear diagrams

To help you extract the important information and decide which technique to use, the first step in answering any question is to draw a diagram. A clear diagram greatly improves your chances of setting up your equations correctly.

Many students attempt to answer questions without drawing a diagram, believing that they are saving time. This is a false economy!

Diagram conventions

To enable you to get all relevant information on your diagram in a structured way:

- Draw your diagram in pencil.
- Whenever a body is moving, include arrows to show the size and direction of velocities and accelerations.
- Then draw any forces in pen so that they are easily distinguished.

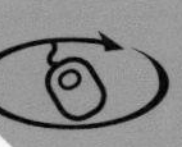

Modelling in mechanics

Mathematical models

In setting up a mathematical model of a real-life situation you have to make some simplifying assumptions.

For example, if you were asked what forces acted on a ball moving through the air, you might well say 'air resistance'. But it is hard to model air resistance, so we often ignore it. We say that the only force acting on the ball is its weight. Fortunately the effect of gravity is much greater than that of air resistance, so our model is simplified without too much loss of accuracy.

That is the key to all modelling assumptions: they concentrate on the most important factors, in order to simplify calculations without too much loss of accuracy.

The process of modelling a problem is best summarised by a diagram:

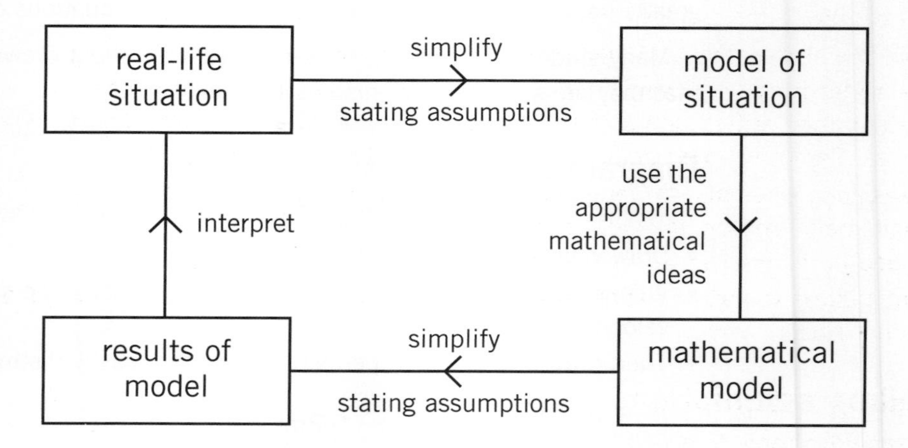

KEY SKILLS

Following the modelling of a real-life problem through the modelling cycle will involve the use of the following skills:

- obtaining and interpreting information
- carrying out multi-task calculations
- evaluating your results
- justifying your choice of methods
- comparison of different methods.

This affords many opportunities to generate evidence for your portfolio for all parts of the Key Skills, Application of Number Level 3.

Glossary of modelling terms

Most of the bodies you meet will be modelled as **particles**. This means that they are small and so may be represented by a single point.

- A **bead** is a particle with a hole drilled through it to enable it to be threaded onto a string or wire.
- A **rod** is an object that is long and thin. Rods are modelled as having length only. Beams, planks, broom handles and so on are all assumed to be rods
- A **lamina** is a flat object whose thickness is small compared with its other dimensions. This page is an example of a lamina. Laminae have area but no volume.
- Laminae and rods are often **uniform**. This means that equal lengths (rods) or areas (laminae) have equal mass. Therefore, by symmetry, the centre of mass is in the middle.
- **Rigid bodies** do not change their shape. The rods that you meet will all be rigid.
- Bodies may be **light**. This means that their mass is small compared with the other masses involved, and so may be ignored.
- **Wires** are always rigid and light.
- **Strings** are always light and **inextensible** (or **inelastic**).
- **Plane surfaces** are completely flat. They may be horizontal or **inclined**. Tables, floors and the Earth's surface, for example, are modelled as horizontal plane surfaces. Ramps or roads going up hills are modelled as inclined planes.
- Surfaces and wires may be **rough** or **smooth**. On a smooth surface, friction is negligible and may be ignored. On a rough surface, a frictional force must be included in the model.
- All the **pulleys** you will meet are smooth. This means that the tension is the same on either side.

Modelling assumptions

You should be able to give a list of assumptions for your model.

A list of assumptions for a cricket ball being thrown might include:

- The ball is a particle.
- The only force acting on the ball is its weight.
- The acceleration due to gravity is constant.
- The ground is a plane horizontal surface.

Vectors 1 – components

Notation

Vectors provide an efficient and elegant way of describing two-dimensional space, or the plane.

The plane may be thought of as a large grid.

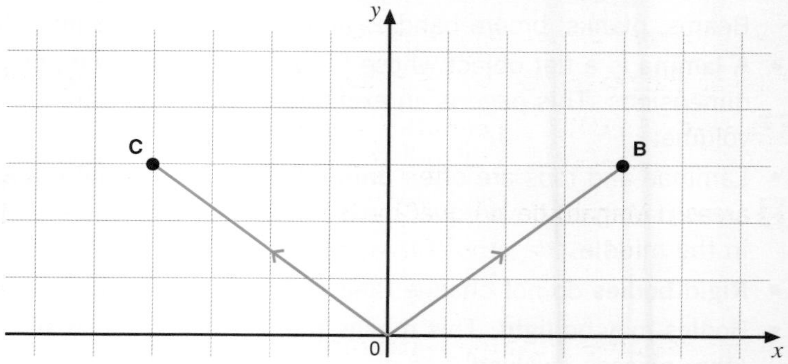

Imagine you could only move along grid lines. So from any point you can only move left, right, up or down.

So to get from O to B (vector OB), you would go right 4 units and then up 3 units.

And to get from O to C (vector OC), you would go left 4 units and then up 3 units.

GURU TIP
You may use either notation in your calculations. But always give the answer in the same notation as the question.

i-j Notation

To simplify the way we describe vectors, two basic unit vectors are defined:

> **i** = movement of one unit to the right
>
> **j** = movement of one unit up

So OB = 4**i** + 3**j**

How do we describe OC? Well, if one unit to the right is given by **i**, then one unit to the left is given by –**i**. (The minus sign tells you to go backwards, facing in the direction of **i**). Similarly, one unit down is given by –**j**.

So OC = –4**i** + 3**j**

This is called **i-j** notation. The general form of a vector in this notation is **a** = X**i** + Y**j**. X and Y are called the 'resolved parts' or 'components' of **a** in the directions of **i** and **j** respectively.

Column-vector notation

Column vectors are an alternative notation. The horizontal component goes at the top. Again, movements to the right and up are taken as positive.

So OB = $\begin{pmatrix} 4 \\ 3 \end{pmatrix}$

and OC = $\begin{pmatrix} -4 \\ 3 \end{pmatrix}$

Throughout this book, **i-j** notation is used, as it is more common in AS courses.

Magnitude, direction and components

So far we have only been moving along the grid lines. The vector has been described in terms of its components. We could, alternatively, describe it in terms of its 'magnitude' and 'direction'. This means giving the length of the vector and its angle from the x axis (that is, from the unit vector **i**).

The magnitude of a vector **a** is written like this: $|\mathbf{a}|$.

Try some examples to get a feel for this topic.

Finding the magnitude and direction of a vector in component form

EXAMPLE Find the magnitude and direction of the vector $\mathbf{a} = 2\mathbf{i} + 3\mathbf{j}$.

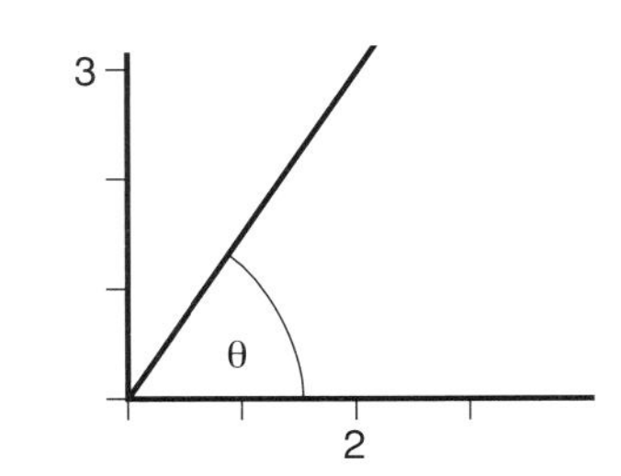

SOLUTION Magnitude $|\mathbf{a}| = \sqrt{(2^2 + 3^2)} = \sqrt{13}$ (this is just Pythagoras' Theorem)

To find the angle from the x axis (and this is your old friend trigonometry):

$$\tan\theta = \frac{3}{2} \Rightarrow \theta = \tan^{-1}\frac{3}{2}$$

$$\Rightarrow \theta = 56° \text{ to the nearest degree}$$

You also need to be able to do the opposite: find the components of a vector given its magnitude and direction.

Resolving a vector into perpendicular components

EXAMPLE A force of 50N acts at an angle of 60° to the Ox axis as shown.

Find the components of the force in the Ox and Oy directions.

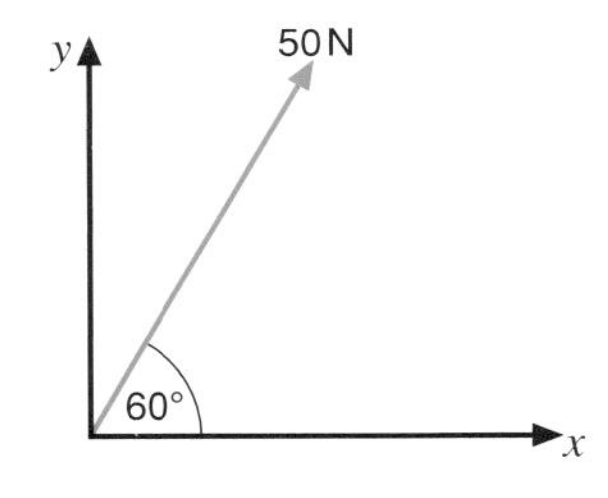

SOLUTION Draw a right-angled triangle and use trigonometry:

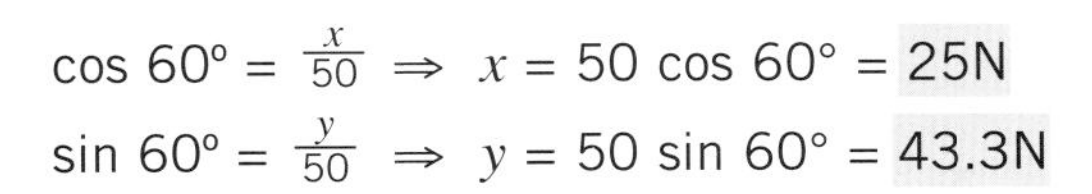

$$\cos 60° = \frac{x}{50} \Rightarrow x = 50\cos 60° = 25\text{N}$$

$$\sin 60° = \frac{y}{50} \Rightarrow y = 50\sin 60° = 43.3\text{N}$$

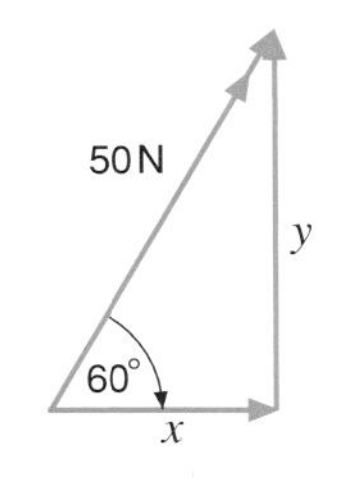

GURU TIP
To close the angle use cos.

Relationship between a vector and its components

Let's look at the general picture now.

To find magnitude and direction:

$|\mathbf{r}| = \sqrt{(x^2 + y^2)}$

$\tan\theta = \frac{y}{x}$

To resolve into components:

$x = |\mathbf{r}| \cos\theta$

$y = |\mathbf{r}| \sin\theta$

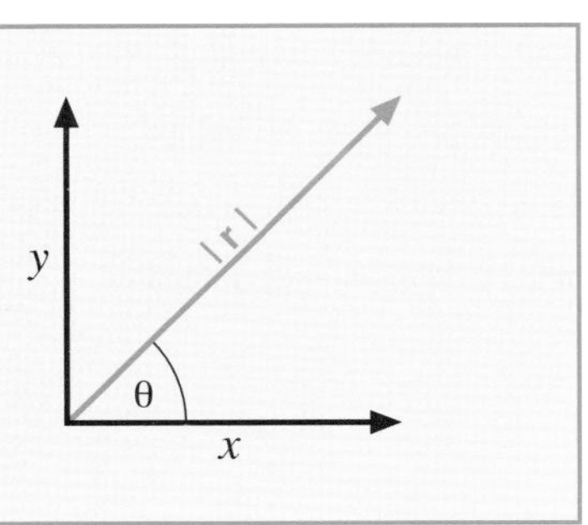

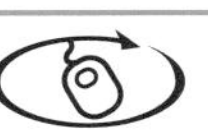

Vectors 2 – vector addition

Adding vectors

To add vectors in **i**-**j** notation, add the **i** and **j** terms separately.

To add vectors in column-vector notation, add the top and bottom numbers separately.

EXAMPLE What is the resultant of the forces (2**i** + 3**j**) N, (7**i** – 10**j**) N and (–8**i** – 6**j**) N?

SOLUTION 2**i** + 7**i** – 8**i** = **i**, and 3**j** – 10**j** – 6**j** = –13**j**

So the resultant is (**i** – 13**j**) N

Vectors may also be added geometrically by drawing them end to end in turn. Draw the first vector. Then, from the end, or head, of it, draw the next. Keep going until you have drawn all the vectors. The result of drawing the vectors 'head to tail' will be a 'polygon of vectors'. The resultant is the vector from the starting point to the end point: that is, from the tail of the first to the head of the last vector. When drawing diagrams, use a double arrow to show the resultant.

EXAMPLE A boat has a cruising speed of $8ms^{-1}$. The current is north-east at $6ms^{-1}$.

If the boat is to travel on a bearing of 090° (due east), what bearing should it steer?

Find also the resultant velocity of the boat.

GURU TIP
Draw a clear diagram, marking on all the information you are given.

SOLUTION

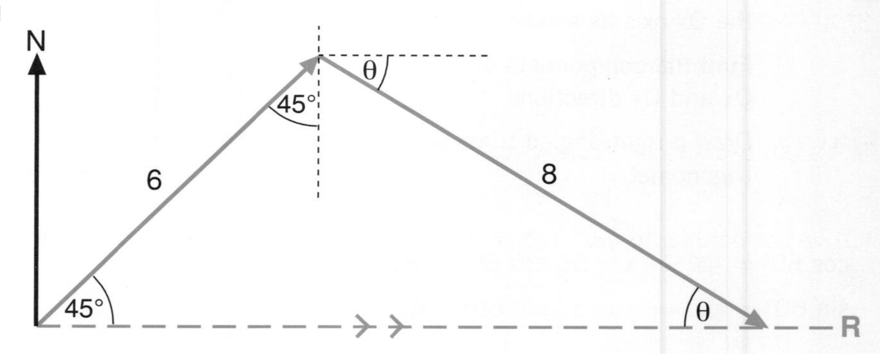

This is an example of a velocity diagram.

Using the sine rule,

$$\frac{\sin 45°}{8} = \frac{\sin\theta}{6} \Rightarrow \sin\theta = 6 \times \frac{\sin 45°}{8} \Rightarrow \theta = \sin^{-1} 0.530 = 32°$$

So the required bearing is 90° + 32° = 122°

The resultant velocity is given by:

$$\frac{|\mathbf{R}|}{\sin(45° + (90° - 32°))} = \frac{8}{\sin 45°}$$

$$\Rightarrow |\mathbf{R}| = 8 \times \frac{\sin 103°}{\sin 45°} = 11.0$$

The boat's speed is $11.0ms^{-1}$

KEY SKILLS
Use of vectors in a practical situation is evidence for Key Skills N3.2 (see the chart on page 6).

Unit vectors

EXAMPLE Find a unit vector in the direction of vector $\mathbf{a} = 5\mathbf{i} - 12\mathbf{j}$.

SOLUTION 1. First find the magnitude of the vector:

$|\mathbf{a}| = \sqrt{(5^2 + 12^2)} = \sqrt{169} = 13$

2. Next, divide the vector by its magnitude to get a unit vector:

Unit vector $= \frac{1}{13}(5\mathbf{i} - 12\mathbf{j})$

> **GURU TIP**
> The final answers to calculations in this book are highlighted in green. In your work, you should highlight your answers, too – a double underline works well.

EXAMPLE Kali the chicken runs at a speed of 3ms^{-1} in the direction of vector $3\mathbf{i} - 4\mathbf{j}$. Write down Kali's velocity vector in component form.

SOLUTION As speed is the magnitude of a velocity vector, we require a vector in the direction of $3\mathbf{i} - 4\mathbf{j}$ with magnitude 3.

1. First find a unit vector in the direction of $3\mathbf{i} - 4\mathbf{j}$:

$|3\mathbf{i} - 4\mathbf{j}| = \sqrt{(3^2 + 4^2)} = \sqrt{25} = 5$

So unit vector $= \frac{1}{5}(3\mathbf{i} - 4\mathbf{j})$

2. Next, multiply the unit vector by the speed to get the velocity vector:

Velocity vector $= \frac{3}{5}(3\mathbf{i} - 4\mathbf{j})\ \text{ms}^{-1}$

Parallel vectors

Two vectors are parallel if the ratios of their components are equal.

EXAMPLE Vectors $\mathbf{p} = a\mathbf{i} - 12\mathbf{j}$ and $\mathbf{q} = 3\mathbf{i} + 4\mathbf{j}$ are parallel. Find the value of a.

SOLUTION For the vectors to be parallel,

$\frac{a}{3} = \frac{12}{4} \Rightarrow \frac{a}{3} = 3 \Rightarrow a = 9$

There are two special cases of this:

- A vector parallel to the $\mathbf{i}$ vector is $k\mathbf{i}$. It has no $\mathbf{j}$ component.
- A vector parallel to the $\mathbf{j}$ vector is $k\mathbf{j}$. It has no $\mathbf{i}$ component.

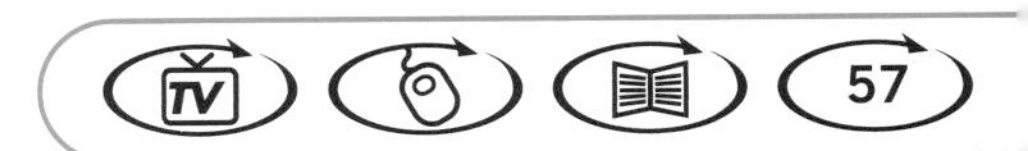

Vectors 3 – vectors in mechanics

The quantities we study in mechanics can be divided into two types: **scalar** and **vector**.

A scalar quantity has only **magnitude**, or size, and is described by a number.

A vector quantity has both magnitude and **direction**.

Some key scalar and vector quantities are:

- Scalar quantities: distance, speed, mass, time.
- Vector quantities: displacement, position, velocity, acceleration.

There are certain relationships between these.

Displacement, position and distance

Let's begin in one dimension. Imagine three towns on a straight railway line that runs east-west:

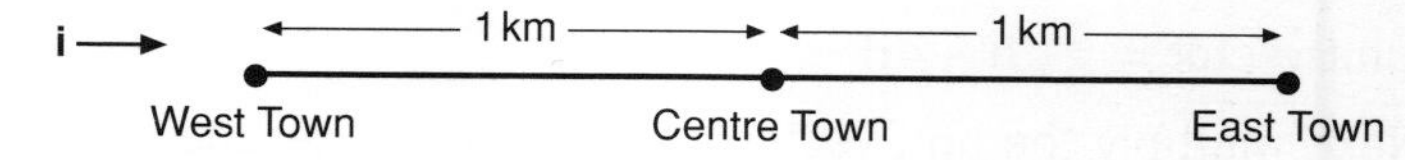

East Town and West Town are both 1km from Centre Town, but, they are in opposite directions. If we call east the positive direction, with a unit vector **i**, then west is negative. So East Town is **i** km from Centre Town and West Town is –**i** km from Centre Town. These are **displacements**. So the displacement of West Town from East Town is –2**i** km.

Now in two dimensions:

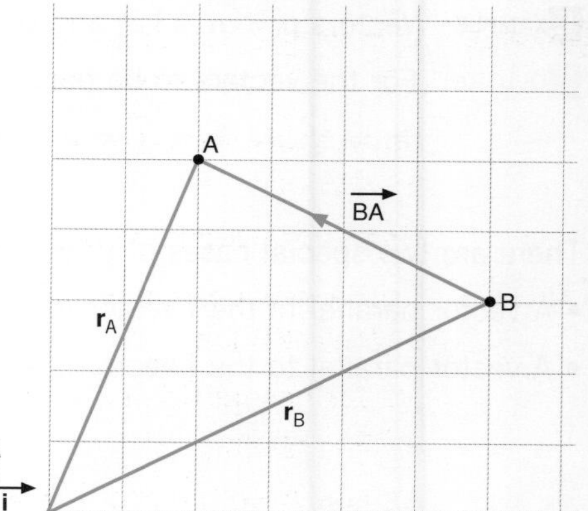

The displacement of A from (or 'relative to') B is

$\overrightarrow{BA} = -4\mathbf{i} + 2\mathbf{j}$

Displacements from some fixed position, or origin, are called position vectors. They are generally labelled **r**:

$\mathbf{r}_A = 2\mathbf{i} + 5\mathbf{j}$

$\mathbf{r}_B = 6\mathbf{i} + 3\mathbf{j}$

A vector triangle gives

$\overrightarrow{BA} = \mathbf{r}_A - \mathbf{r}_B = (2\mathbf{i} + 5\mathbf{j}) - (6\mathbf{i} + 3\mathbf{j}) = -4\mathbf{i} + 2\mathbf{j}$

In general this may be written:

> displacement = change in position
> = new position – original position

Rearranging gives:

> new position = original position + displacement

Speed and velocity

Let's return to our one-dimensional world of three towns. Imagine a train going from West Town to East Town and back, travelling at a constant speed of 25ms^{-1}. The station-master at Centre Town will see the train pass first travelling at 25 ms^{-1} going east and then at 25ms^{-1} going west. In terms of the unit vector defined above, these are 25$\mathbf{i}$ ms^{-1} and $-25\mathbf{i}$ ms^{-1}. These vector quantities are **velocities**.

Velocity vectors extend into two dimensions in a similar way to displacement vectors.

Speed is the magnitude of a velocity vector.

Position, displacement and velocity

From GCSE you may remember the formula 'distance = speed × time', which applies when the speed is constant.

The vector equivalent of this formula is:

displacement = velocity × time

Rearranging gives:

$$\text{velocity} = \frac{\text{displacement}}{\text{time}}$$

$$\text{velocity} = \frac{\text{change in position}}{\text{change in time}}$$

Another useful equation is:

new position = original position + velocity × time

EXAMPLE At midnight, two ships A and B are at positions $(4\mathbf{i} + 6\mathbf{j})$ km and $(-3\mathbf{i} + 2\mathbf{j})$ km relative to a third ship.

The velocity vectors of the two ships are $(\mathbf{i} - 2\mathbf{j})$ kmh^{-1} and $(\mathbf{i} - \mathbf{j})$ kmh^{-1} respectively.

Find their distance apart at 6am.

SOLUTION Ship A new position $= (4\mathbf{i} + 6\mathbf{j}) + (\mathbf{i} - 2\mathbf{j}) \times 6 = (4\mathbf{i} + 6\mathbf{j}) + (6\mathbf{i} - 12\mathbf{j})$
$= (10\mathbf{i} - 6\mathbf{j})$ km

Ship B new position $= (-3\mathbf{i} + 2\mathbf{j}) + (\mathbf{i} - \mathbf{j}) \times 6 = (-3\mathbf{i} + 2\mathbf{j}) + (6\mathbf{i} - 6\mathbf{j})$
$= (3\mathbf{i} - 4\mathbf{j})$ km

We now have the position vectors of the two ships. The distance apart is the magnitude of the displacement vector.

displacement $\overrightarrow{AB}$ = position of B – position of A

$= (3\mathbf{i} - 4\mathbf{j}) - (10\mathbf{i} - 6\mathbf{j}) = (-7\mathbf{i} + 2\mathbf{j})$ km

So distance AB $= |\overrightarrow{AB}| = \sqrt{(7^2 + 2^2)} = \sqrt{53} =$ 7.28 km

Acceleration

When the speed of a body is changing, the body is said to be accelerating.

Acceleration is the rate of change of velocity. If the acceleration is constant, then

$$\text{acceleration} = \frac{\text{final velocity} - \text{initial velocity}}{\text{final time} - \text{initial time}}$$

You may have noticed some glaring omissions in the list of vector quantities at the beginning of this section. **Force** can be defined in terms of acceleration, and is a vector quantity. **Impulse** and **momentum** can be defined in terms of velocity, and are also vector quantities.

Kinematics 1 – constant acceleration

This is just an extension of the work you did on distance-time and velocity-time graphs at GCSE.

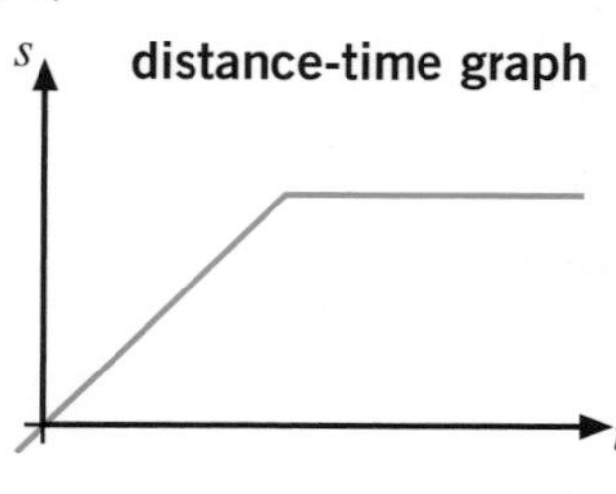

The gradient of the graph is the velocity.

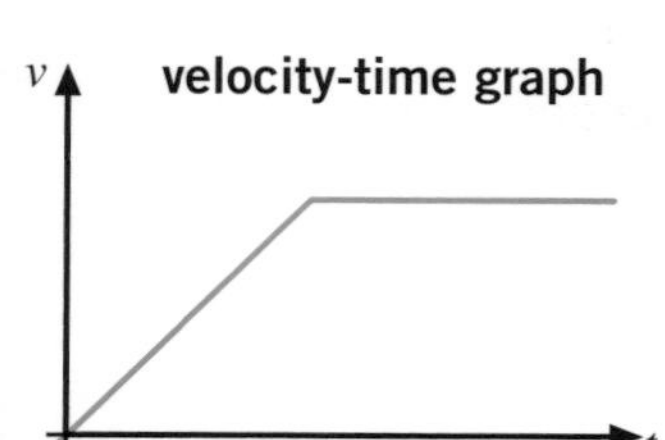

The gradient of the graph is the acceleration.

The area under the graph is the distance travelled.

Equations of motion for constant acceleration

You will need to know and be able to use the following five equations:

	where	
$v = u + at$		u = initial velocity
$s = ut + \frac{1}{2}at^2$		v = final velocity
$v^2 = u^2 + 2as$		a = acceleration
$s = \frac{(u + v)\,t}{2}$		s = displacement
$s = vt - \frac{1}{2}at^2$		t = time taken

These can all be derived from the following diagram.

Try doing this. If you get stuck, ask your teacher.

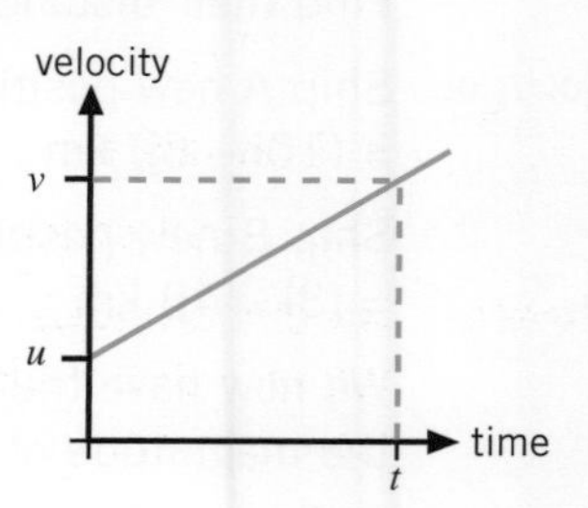

EXAMPLE A train accelerating uniformly in a straight line passes a point A with a speed of 9ms^{-1}. 20s later it passes a point B with a speed of 21ms^{-1}. How far apart are A and B, and what is the train's acceleration?

SOLUTION 1. First, draw a diagram.

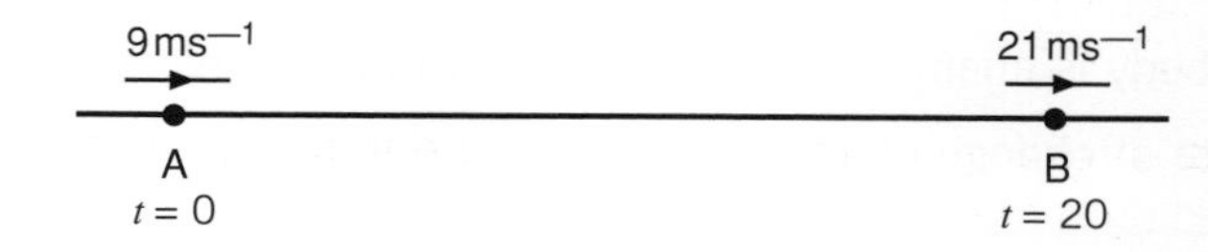

2. Next, write down the five key variables and any values that you know.

u	v	s	a	t
9	21			20

3. Now select the equation to use. Each equation links four of the five variables. Choose the one that links the three you are given with the one you want.

To find the distance from A to B, we need to find s given u, v and t.

Use $s = \frac{(u + v)\,t}{2} \Rightarrow s = \frac{(9 + 21) \times 20}{2} \Rightarrow s = 300\text{m}$

To find the train's acceleration, you need to find a given u, v and t.

Use $v = u + at \Rightarrow 21 = 9 + 20a \Rightarrow 12 = 20a \Rightarrow a = 0.6 \text{ ms}^{-2}$

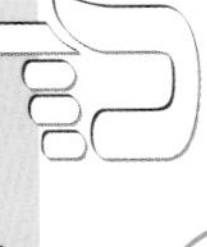

GURU TIP
Even once you've calculated s, avoid using this in case you made a mistake in your calculation.

Vertical motion

Exam questions often involve objects dropped or thrown vertically.

When an object is dropped vertically, remember that the initial velocity is 0.

When an object is thrown vertically, remember:

- The final velocity at the top is 0.
- The total time in the air is twice the time to the top.
- When the particle returns to its starting point, the displacement $s = 0$.

In all of these problems, the acceleration downwards is g: the acceleration due to gravity. This may be taken as 10ms^{-2}, 9.8ms^{-2} or 9.81ms^{-2} depending on the exam board.

GURU TIP
Confirm with your teacher which value you should use.

For the examples in this book, we will take $g = 9.8 \text{ ms}^{-2}$.

EXAMPLE A man drops a cannonball from the top of a tower. The canonball takes 10s to hit the ground. How high is the tower?

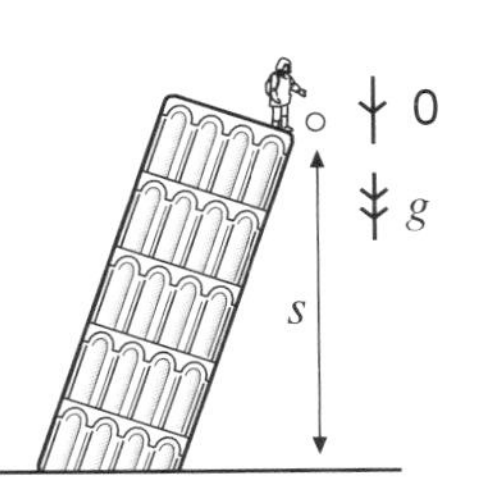

SOLUTION 1. First, draw a diagram.

2. Next write out the variables and identify which direction is positive.

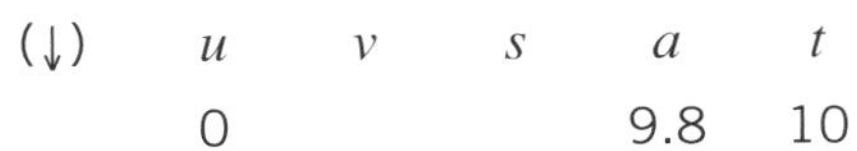

(↓)	u	v	s	a	t
	0			9.8	10

GURU TIP
Always show which direction you are taking as positive.

3. Now use the equation: $s = ut + \frac{1}{2}at^2 \Rightarrow s = \frac{1}{2} \times 9.8 \times 100 = 490$

The height of the tower is 490m.

EXAMPLE A woman throws a cannonball vertically up from the top of a 343m-high cliff, with a velocity of 14.7ms^{-1}. How long will it take the cannonball to hit the sea?

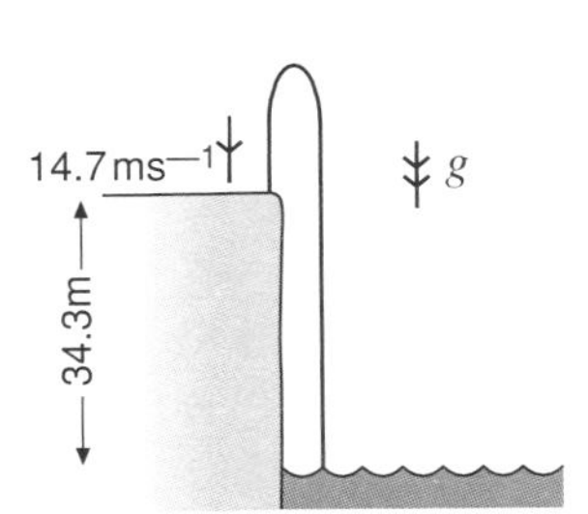

SOLUTION 1. Draw a diagram.

2. Write out the variables.

(↓)	u	v	s	a	t
	−14.7		242	9.8	

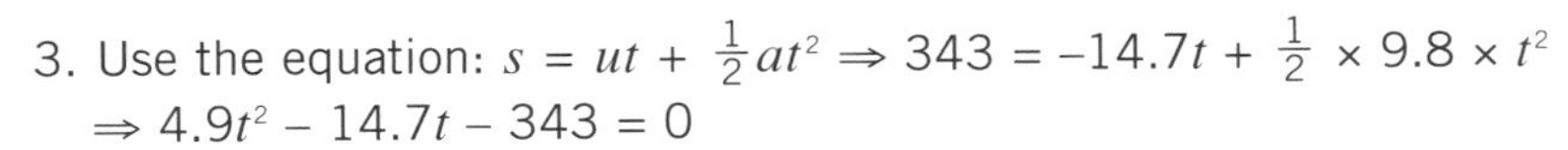

3. Use the equation: $s = ut + \frac{1}{2}at^2 \Rightarrow 343 = -14.7t + \frac{1}{2} \times 9.8 \times t^2$

$\Rightarrow 4.9t^2 - 14.7t - 343 = 0$

You need to solve a quadratic equation. The numbers may look horrible, but try dividing through by the t^2 coefficient. In most cases the equation will then factorise.

$t^2 - 3t - 70 = 0 \Rightarrow (t + 7)(t - 10) = 0 \Rightarrow t = 10$ or -7

The equation has two solutions, one positive and one negative. The answer to the question must be positive.

So the time to hit the sea is 10 seconds.

Mechanics

Kinematics 2 – variable acceleration

If the acceleration is not constant, you cannot use the five equations of motion.

Velocity is the rate of change of position:

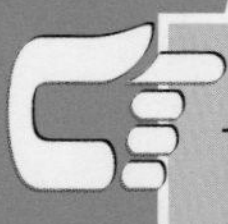
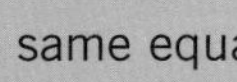

GURU TIP
The same equations apply when you are talking about vector quantities.

$$v = \frac{dr}{dt}$$

Acceleration is the rate of change of velocity:

$$a = \frac{dv}{dt}$$

Fortunately, in exams the functions chosen are nice easy ones.

In pure mathematics, most equations express y as a function of x: for example, $y = 6x^2 + 2x$. In kinematics, most equations will express a, v or r as a function of t, as they are telling us something about a particle's acceleration, velocity or position at a given time. The letters may be different but the rules of differentiation are the same.

Remember the following way of moving from one equation to another. Don't forget to add a constant when integrating. The question will contain information to enable you to evaluate the constant.

GURU TIP
If the question asks for an equation giving information about the force F, find an equation for a and use $F = ma$. See page 68.

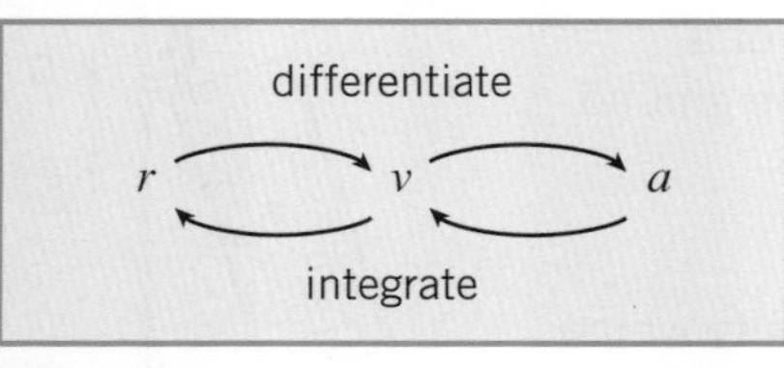

EXAMPLE The velocity v ms^{-1} of a particle moving in a straight line at a time t seconds after starting to move is given by $v = 10 + 2t + t^2$.

(a) Find an equation for the acceleration a ms^{-2} in terms of t.

(b) What is the initial acceleration?

(c) Find an equation for the displacement in terms of t.

SOLUTION (a) To find the equation for acceleration, differentiate the equation for the velocity:

$v = 10 + 2t + 3t^2 \Rightarrow a = \frac{dv}{dt} = 2 + 6t$

(b) The initial acceleration is given by substituting $t = 0$ into the above equation:

$t = 0$, $a = 2\text{ms}^{-2}$

(Include the '$t = 0$' to show the examiner what you are doing.)

(c) To find the equation for displacement, integrate the equation for the velocity, not forgetting to add the constant of integration:

$v = 10 + 2t + 3t^2 \Rightarrow r = \int v\,dt = 10t + t^2 + t^3 + c$

Since r is the displacement from the starting point, when $t = 0$, $r = 0$.

$t = 0$, $r = 0 \Rightarrow c = 0$.

So the equation is:

$r = 10t + t^2 + t^3$

Questions on this topic often involve vector notation.

EXAMPLE A particle P of mass 0.5kg moves in a plane. At time t seconds, its velocity $\mathbf{v}$ ms^{-1} is given by

$\mathbf{v} = (3t^2 - 1)\,\mathbf{i} + 8t^3\,\mathbf{j},\ t \geq 0.$

At time $t = 0$, the position vector of P relative to a fixed origin O is given by $3\mathbf{j}$ m.

When $t = 5$, P is at the point B.

Find:

(a) the position vector of B

(b) the magnitude of the force on P at B

SOLUTION As we are starting with a velocity equation and need information about the position at a given time, the first thing to do is to integrate the velocity equation. To do this, integrate each part separately, remembering to add constants:

$\mathbf{r} = \int \mathbf{v}\,dt = \int ((3t^2 - 1)\,\mathbf{i} + 8t^3\,\mathbf{j})\,dt = (t^3 - t + c)\,\mathbf{i} + (2t^4 + k)\,\mathbf{j}$

Now use the initial conditions (the information in the question about $t = 0$ and $t = 5$) to evaluate the constants:

$t = 0,\ \mathbf{r} = 3\mathbf{j} \Rightarrow 3\mathbf{j} = (0^3 - 0 + c)\,\mathbf{i} + (2 \times 0^4 + k)\,\mathbf{j} = c\mathbf{i} + k\mathbf{j}$

Just split this into two equations – one for the **i** components and one for the **j** components:

$c\mathbf{i} = 0\mathbf{i} \Rightarrow c = 0$

$k\mathbf{j} = 3\mathbf{j} \Rightarrow k = 3$

So the equation is:

$\mathbf{r} = (t^3 - t)\,\mathbf{i} + (2t^4 + 3)\,\mathbf{j}$

This gives you the position vector of P at any time t. So to find the position vector of B, just substitute in the time when P arrives there:

$t = 5,\ \mathbf{r}_B = (5^3 - 5)\,\mathbf{i} + (2 \times 5^4 + 3)\,\mathbf{j} = (120\mathbf{i} + 1253\mathbf{j})$ m

Integration is more difficult than differentiation. You need to remember to add constants. To answer part (b), we need to find an equation telling us about the force at time t. We want an equation for **a**. So we differentiate the velocity equation:

$\mathbf{a} = \frac{d\mathbf{v}}{dt} = 6t\,\mathbf{i} + 24t^2\,\mathbf{j}$

Using $\mathbf{F} = m\mathbf{a}$:

$\mathbf{F} = 0.5 \times (6t\,\mathbf{i} + 24t^2\,\mathbf{j}) = 3t\,\mathbf{i} + 12t^2\,\mathbf{j}$

This gives us the force applied at any time t. So at B,

$t = 5 \Rightarrow \mathbf{F} = 3 \times 5\,\mathbf{i} + 12 \times 5^2\,\mathbf{j} = 15\mathbf{i} + 300\mathbf{j} \Rightarrow |\mathbf{F}| = \sqrt{(15^2 + 300^2)}$
$= 300.4$ N

GURU TIP
The equation $\mathbf{F} = m\mathbf{a}$ is discussed on pages 70-71.

Projectiles 1 – equations of motion

A **projectile** flies through the air under the action of gravity alone.

Treat the horizontal and vertical motions separately, and use the equations of motion.

You need to resolve the initial velocity into horizontal and vertical components.

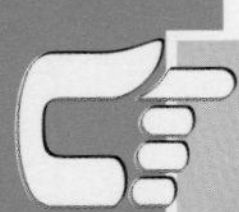

GURU TIP
Check with your teacher whether your exam board requires you to know this.

In general, the components are:

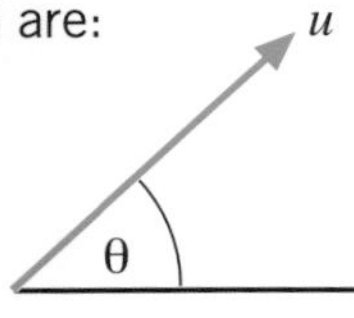

≡

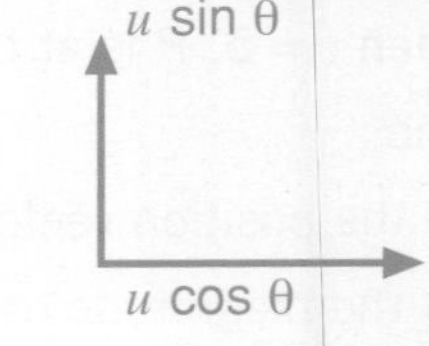

So you have the following information:

- Horizontally there is no acceleration.
- Vertically the acceleration is that due to gravity and is downwards.
- The times in the vertical and horizontal motions are the same.

↑	u	v	s	a	t
	10			–9.8	

→	u	$v\ (=u)$	s	a	t
	17.3			0	

Finding the greatest height

At the top of the trajectory, $v\uparrow = 0$.

Continuing with the above example, and looking at the vertical motion:

↑	u	v	s	a	t
	10	0		–9.8	

We are given u, v and a and wish to find s.

Use: $v^2 = u^2 + 2as$

$v^2 = u^2 + 2as \Rightarrow 0^2 = 10^2 + 2 \times (-9.8) \times s \Rightarrow 100 = 19.6s \Rightarrow s = 5.10$

The greatest height of the particle is 5.10m.

Finding the time of flight and range on horizontal ground

To find out for how long the projectile is in the air, consider the vertical motion.

When the projectile hits the ground, $s\uparrow = 0$.

The range is then the horizontal distance travelled in this time.

Continuing with the example above:

↑	u	v	s	a	t
	10		0	–9.8	

We are given u, s and a and wish to find t.

Use: $s = ut + \frac{1}{2}at^2$

$s = ut + \frac{1}{2}at^2 \Rightarrow 0 = 10t + \frac{1}{2} \times (-9.8)t^2 \Rightarrow 0 = t\,(10 - 4.9\,t) \Rightarrow t = 0$ or 2.04

We want the positive solution (the other solution just tells us that when $t = 0$, $s\uparrow = 0$).

The time of flight of the particle is $2.04s$.

Now consider the horizontal motion:

$\rightarrow$	u	$v\ (=u)$	s	a	t
	17.3			0	2.04

Use the equation: $s = ut + \frac{1}{2}at^2 \Rightarrow s = 17.3 \times 2.04 = 35.3$

The horizontal distance travelled is 35.3m.

GURU TIP
You can ignore the 'trivial' solution $t = 0$.

Particle projected from a high place

EXAMPLE A boy standing at the top of a 171.5m-high cliff throws a ball with a velocity of 19.6ms^{-1} at an angle of 30° to the horizontal.

(a) Find the greatest height of the ball above the sea.

(b) Show that the time t seconds taken to hit the sea is given by $t^2 - 2t - 35 = 0$.

(c) Solve this to find t.

SOLUTION (a) Find the horizontal and vertical components of the initial velocity:

$u\rightarrow = 19.6 \cos 30° = 17.0$

$u\uparrow = 19.6 \sin 30° = 9.8$

At the highest point, $v\uparrow = 0$.

$\uparrow$	u	v	s	a	t
	9.8	0		−9.8	

Use the equation: $v^2 = u^2 + 2as \Rightarrow 0^2 = 9.8^2 + 2 \times (-9.8) \times s \Rightarrow s = 4.9$

As this is the height above the cliff, the greatest height above the ground is $171.5 + 4.9 =$ 176.4m.

(b) 'Up' has been taken as the positive direction. Taking the top of the cliff as the origin, the distance to the ground is −171.5m.

$\uparrow$	u	v	s	a	t
	9.8		−171.5	−9.8	

Use the equation: $s = ut + \frac{1}{2}at^2 \Rightarrow -171.5 = 9.8\,t + \frac{1}{2} \times (-9.8)t^2 \Rightarrow$
$-171.5 = 9.8\,t - 4.9\,t^2 \Rightarrow 4.9\,t^2 - 9.8\,t - 171.5 = 0$

This doesn't look much like the equation we're supposed to have, but dividing throughout by 4.9 gives:

$t^2 - 2t - 35 = 0$

(c) $t^2 - 2t - 35 = 0 \Rightarrow (t - 7)(t + 5) = 0 \Rightarrow t = 7$ or $-5 \Rightarrow t = 7$ (taking the positive value).

The ball hits the sea after 7 seconds.

Examiners sometimes ask you to find the time for which a projectile is above a certain height, say h. To answer this, use $h = ut + \frac{1}{2}at^2$ and then solve the quadratic. This will give you the two times at which the projectile is at height h. In between, it must be above that height. So just subtract the smaller solution from the larger one.

Projectiles 2 – the general case

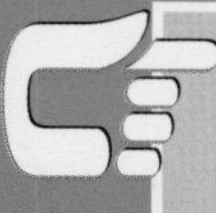

GURU TIP
The path of a projectile is a parabola. Its equation is a quadratic.

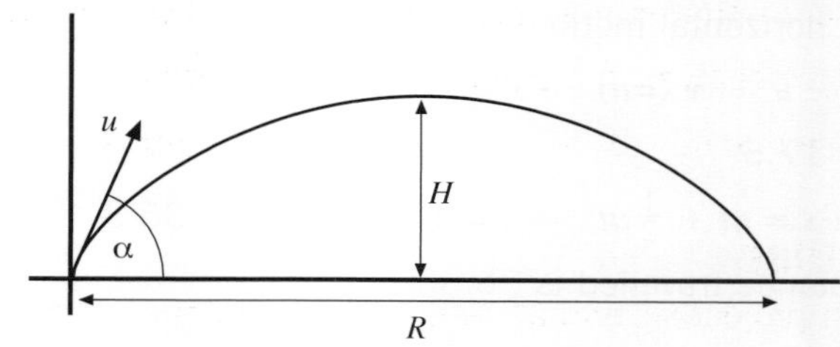

To find the equation of the path of a projectile:

1. Find equations in terms of t for the horizontal (x) and vertical (y) components of the position.
2. Eliminate t by substitution to obtain the equation of the path.

The general equation of the path can then be used to determine whether points are accessible by the projectile.

EXAMPLE A projectile is launched from the origin with an initial speed of 30ms^{-1} at an angle of 60° to the horizontal.

(a) Find the position of the particle after t seconds.

(b) Find the equation of the path.

(c) Does the projectile reach a height of 35 metres?

SOLUTION Find the horizontal and vertical components of the velocity:

$v\uparrow = 30 \cos 60° = 15$

$v\rightarrow = 30 \sin 60° = 26$

(a) Consider the horizontal motion:

→	u	$v\ (=u)$	s	a	t
	15	15	x	0	t

So $x = 15\,t$ (1)

Consider the vertical motion:

↑	u	v	s	a	t
	26		y	−9.8	t

So $y = 26\,t - 4.9\,t^2$ (2)

The position after t seconds is given by:

$x = 15\,t$

$y = 26\,t - 4.9\,t^2$

(b) Rearranging equation (1) gives

$t = \frac{x}{15}$

Substituting this into equation (2) gives:

$y = 26 \times \frac{x}{15} - 4.9 \times (\frac{x}{15})^2$

$= 1.73\,x - 0.022\,x^2$

(c) $y = 35 \Rightarrow 35 = 1.73\,x - 0.022\,x^2 \Rightarrow 0.022\,x^2 - 1.73\,x + 35 = 0$

This is a quadratic equation: $ax^2 + bx + c = 0$
where $a = 0.022$, $b = -1.73$ and $c = 35$.

The discriminant is $b^2 - 4ac = 1.73^2 - 4 \times 0.022 \times 35 = -0.087$

As this is negative, the equation has no solution.

The projectile cannot reach a height of 35 metres.

General formulae

Some questions on projectiles will ask you to find an equation for the general case using algebra. These questions may look scary, but there are really only a few things that examiners can ask...

GURU TIP
Make sure you understand these derivations rather than just memorising the results.

Mechanics

Finding the maximum height

At the greatest height, $v\uparrow = 0$

So using $v = u + at$ and considering the vertical motion,

$$0 = u \sin \alpha - gt \Rightarrow u \sin \alpha = gt \Rightarrow t = \frac{u \sin \alpha}{g}$$

Substitute this into $s = ut + \frac{1}{2} at^2$ and call the maximum height H:

$$H = (u \sin \alpha) \frac{u \sin \alpha}{g} - \frac{1}{2} g \left(\frac{u \sin \alpha}{g}\right)^2$$

$$= \frac{u^2 \sin^2 \alpha}{g} - \frac{1}{2} g \frac{u^2 \sin^2 \alpha}{g^2}$$

$$= \frac{u^2 \sin^2 \alpha}{g} - \frac{u^2 \sin^2 \alpha}{2g}$$

(a factor of g cancels in the second term)

$$H = \frac{u^2 \sin^2 \alpha}{2g}$$

Finding the time of flight

When the projectile hits the ground (as long as the ground is horizontal), $s\uparrow = 0$.

Consider the vertical motion. Use $s = ut + \frac{1}{2} at^2$:

$0 = (u \sin \alpha) t - \frac{1}{2} gt^2 \Rightarrow 0 = t (u \sin \alpha - \frac{1}{2} gt) \Rightarrow 0 = u \sin \alpha - \frac{1}{2} gt \Rightarrow u \sin \alpha = \frac{1}{2} gt$

(ignoring the trivial solution $t = 0$)

$$t = \frac{2u \sin \alpha}{g}$$

Finding the range

We now know that the time of flight is given by $t = \frac{2u \sin \alpha}{g}$

So the range is just the horizontal distance covered in that time.

Use $s = ut + \frac{1}{2} at^2$. For the horizontal component, $a = 0$. Let the range be R.

$R = u \cos \alpha \times \frac{2u \sin \alpha}{g}$

So:

$$R = \frac{2u^2 \sin \alpha \cos \alpha}{g}$$

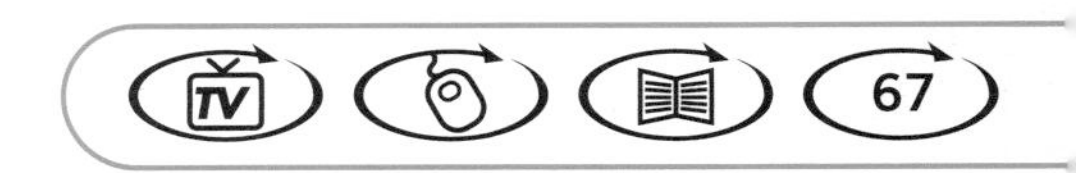

Newton's Laws of Motion

Newton formulated three laws of motion. His Second Law is the one that is most often used explicitly.

GURU TIP
$\sum \mathbf{F}$ stands for the resultant (sum) of the forces.

> Newton's First Law: A particle will remain at rest or continue to move with a constant velocity in a straight line unless acted on by a force.
>
> Newton's Second Law: The force needed produce a given acceleration in a particle of a given mass is proportional to the acceleration and to the mass:
>
> (resultant) force = mass × acceleration
>
> $\sum \mathbf{F} = m\mathbf{a}$
>
> Newton's Third Law: Every action has an equal and opposite reaction.

Using Newton's Second Law in one dimension

Let's start with a nice easy example. The key to all questions in mechanics, particularly those involving forces, is a clear diagram. Draw the diagram in pencil, then add the forces in pen so that they stand out. Remember, forces are vectors and so have direction. The equation $\sum \mathbf{F} = m\mathbf{a}$ is a vector equation. When adding up the forces that ara acting on a particle (to find the **resultant** force), pay attention to the signs. It is usual to take the direction of the acceleration as positive.

EXAMPLE Find the acceleration of a particle of mass 5kg acted on by a resultant force of 20N.

SOLUTION 1. First draw a diagram.

2. Now use the equation: $\sum \mathbf{F} = m\mathbf{a} \Rightarrow 20 = 5a \Rightarrow a = 4$

The acceleration of the particle is 4ms^{-2}.

EXAMPLE A man is cycling along a horizontal road. The combined mass of the man and the bike is 100kg. If the man is exerting a forward force of 200N and the bike is accelerating at 0.5ms^{-2}, find the magnitude of the resisting force.

SOLUTION 1. Draw a diagram with the forces in pen. Show the direction of the acceleration. Let the magnitude of the resisting force be R newtons.

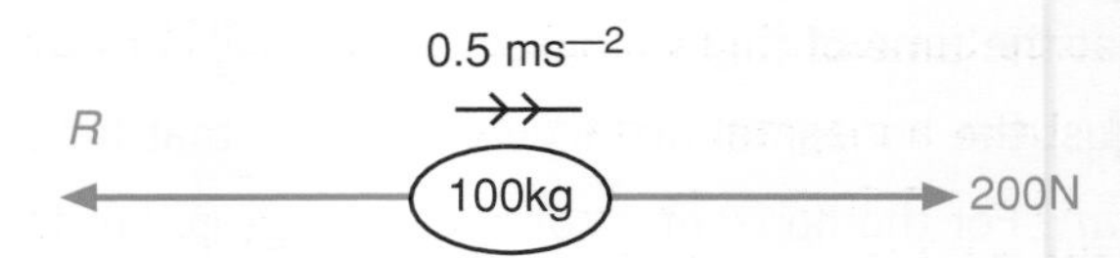

2. Now use the equation $\sum \mathbf{F} = m\mathbf{a}$. Resolve in the direction of the acceleration:

$200 - R = 100 \times 0.5 \Rightarrow 200 - R = 50 \Rightarrow R = 150$

The resisting force is 150N.

Newton's First Law and equilibrium

Newton's First Law tells us that unless a resultant force acts on a body it will remain at rest or moving at a constant velocity.

A body that is at rest or moving at a constant velocity is said to be in **equilibrium.**

For a particle in equilibrium, $\sum \mathbf{F} = \mathbf{0}$.

GURU TIP
Whenever there is contact between an object and a surface there will be a reaction force.

Mechanics

Forces of contact – the normal reaction

Imagine an apple on a table. Suppose the apple has a weight of 1N. Let's look at the force diagram.

Why doesn't the apple move? Because the table is pushing back. But with what force? Well, if it were more than 1N, there would be a resultant force up, and so the apple would move up. If the force were less than 1N, there would be a resultant force down, and so the apple would move down. So the force must be exactly 1N.

This is an example of Newton's Third Law. The table exerts a force upwards on the apple, of the same magnitude as the weight but in the opposite direction. You should include this upwards force on the force diagram.

EXAMPLE A woman of mass 50kg is standing in a lift of mass 200kg, which starts to accelerate upwards at 2.2ms^{-2}.

Find:

(a) the force between the woman and the floor

(b) the tension in the cable

SOLUTION (a) First draw a diagram, labelling the forces. Show the direction of acceleration. We are only interested in the woman and the floor.

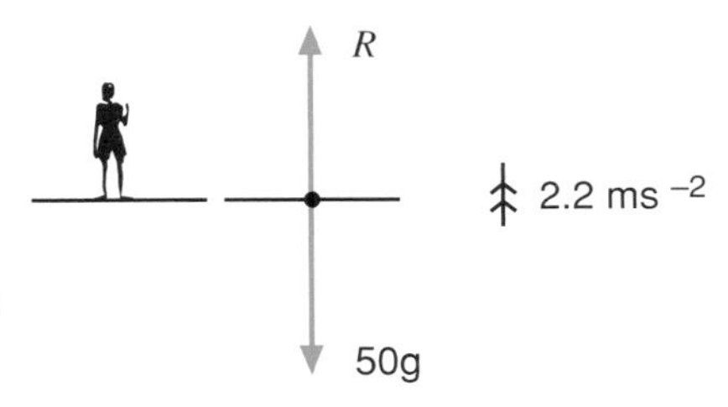

By Newton's Third Law we know that there must be a contact force between the woman and floor. Call it R newtons.

Now use the equation $\sum \mathbf{F} = m\mathbf{a}$. Resolve in the direction of the acceleration:

$R - 50g = 50 \times 2.2 \Rightarrow R = 50 \times 9.8 + 50 \times 2.2$

$= 50 \times 12 = 600$

The force between the woman and the floor is 600N.

(b) 1. Draw a diagram. This time we are just interested in the cable and the total load it carries. Let the tension in the cable be T newtons.

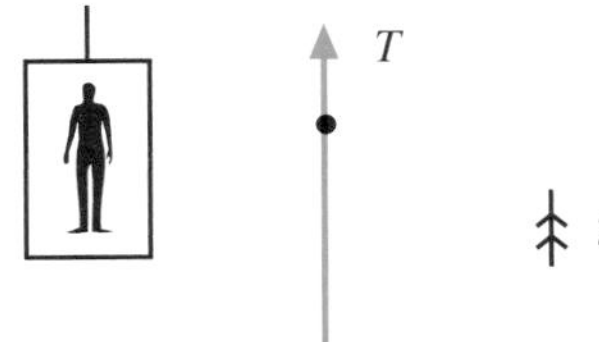

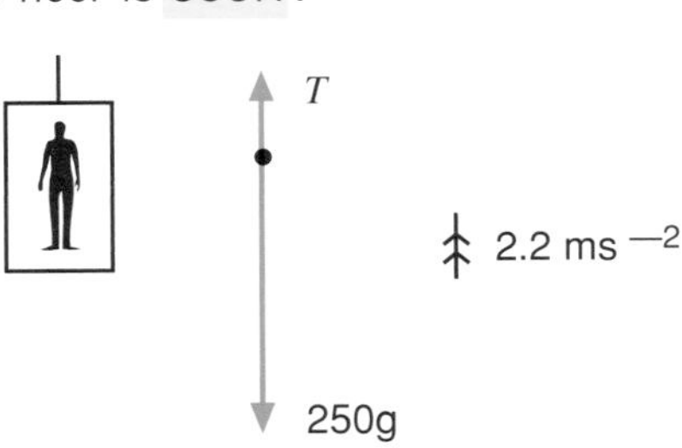

2. Use the equation $\sum \mathbf{F} = m\mathbf{a}$:

$T - 250g = 250 \times 2.2$
$\Rightarrow T = 250 \times (9.8 + 2.2) = 300$

The tension in the cable is 300N.

Connected particles

When two particles are connected by a light inextensible string, there is a force of **tension** in the string.

At any point on the string, there are two forces of tension which are equal and opposite in direction. Include both tensions on your diagrams.

The two particles, being connected, will have the same acceleration. Include this on your diagram.

Then use the equation **F** = m**a** to find equations of motion for the two particles. Take the direction of the acceleration as positive. You will then need to solve a pair of simultaneous equations.

Most exam questions on this topic are either about pulleys or about a vehicle towing a load. Questions involving vehicles usually involve a driving force taking the vehicle forwards and a resisting force in the opposite direction. The tension in the tow bar pulls the load forward but pulls the driving vehicle backward.

EXAMPLE A tractor of mass 1000kg pulls a load of hay of mass 3000kg along a horizontal road. The tractor's engine exerts a forward force of 4kN. The resistance to motion of the tractor and load are proportional to their masses. Given that the tractor accelerates at 0.5ms^{-2}, find the tension in the tow bar.

SOLUTION 1. First, draw a diagram. Include the direction of the acceleration. Let the constant of proportionality between the masses of the tractor and load and their resisting forces be k ms^{-2}. Let the tension in the tow bar be T newtons.

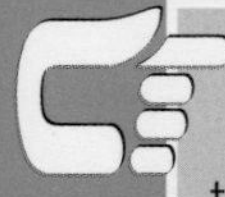

GURU TIP
The tension in the tow bar pulls the load forwards but the tractor backwards.

3000k N | 3000 kg | T | T | 1000 kg | 4000 N | 1000k N

2. Use the equation **F** = m**a** to write equations of motion for the tractor and the load. Resolve the forces in the direction of the acceleration.

For the tractor: $4000 - 1000k - T = 1000 \times 0.5$

$\Rightarrow 4000 - 1000k - T = 500 \Rightarrow T + 1000k = 3500$ (1)

For the load: $T - 3000k = 3000 \times 0.5 \Rightarrow T - 3000k = 1500$ (2)

3. Finally, you have to solve these simultaneous equations:

Equation (1) × 3 gives: $3T + 3000k = 10500$

Adding equation (2) gives: $4T = 12000 \Rightarrow T = 3000$

The tension in the tow bar is 3000N.

Don't be put off by the simultaneous equations – they are quite friendly. They tend to be even more friendly in pulley problems: just add the equations and the tension melts away!

EXAMPLE Two particles A and B of mass 5kg and 4kg respectively are attached to the ends of a light inextensible string which passes over a smooth pulley. The system is released from rest. Find the acceleration of the system and the reaction of the pulley on its support.

SOLUTION As the particles start from rest, they will move in the direction of the larger mass.

1. First, draw a diagram. Because the pulley is smooth, the tension on either side is the same. Let this tension be T newtons, let the reaction of

the pulley on its support be R newtons and let the acceleration be a ms^{-2}.

2. Now write equations of motion for the two particles, showing the direction taken as positive:

A(↓): $5g - T = 5a$

B(↑): $T - 4g = 4a$

3. Eliminate T by adding the two equations:

$g = 9a \Rightarrow a = \frac{9.8}{9} = 1.09$

The particles accelerate at $1.09ms^{-2}$ with A moving downwards.

4. To find the reaction, you will need to write an equation of motion for the pulley. This is always the same. Although the pulley is turning, it is not moving up or down, so its acceleration is zero.

P(↑): $R - 2T = 0 \Rightarrow R = 2T$

To find T, substitute $a = \frac{g}{9}$ into the equation of motion for B:

$T - 4g = 4 \times \frac{g}{9} \Rightarrow T = 4 \times 9.8 + \frac{4}{9} \times 9.8 = 43.56$

$\Rightarrow R = 87.1$

The reaction of the pulley on its support is 87.1N.

GURU TIP
The heavier mass always accelerates downwards, the lighter mass upwards.

Pulley on a smooth inclined plane

EXAMPLE A particle of mass m rests on a smooth plane inclined at an angle of 30° to the horizontal. The particle is attached by a light inextensible string which passes over a smooth pulley to a particle of mass $5m$. The string is taut and the particles are released from rest.

Find the tension in the string the acceleration of the particles.

SOLUTION Draw a diagram. Let the particle of mass m be A. Let the particle of mass $5m$ be B. Let the acceleration of the particles be a ms^{-2}. Let the tension in the string be T newtons.

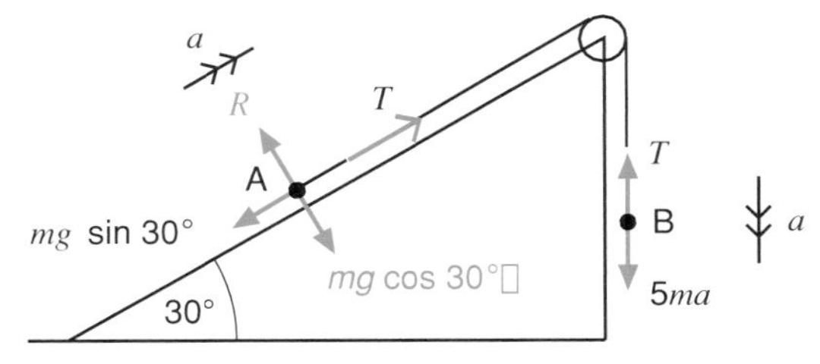

Use the components of the weight of A parallel and perpendicular to the plane.

Use the equation **F**= m**a** to write equations of motion for the two particles:

(A↗): $T - mg \sin 30° = ma$

$\Rightarrow T - \frac{1}{2} mg = ma$ (1)

(B↓): $5mg - T = 5ma$ (2)

Adding equations (1) and (2) gives: $\frac{11}{2} mg = 6ma$

$\Rightarrow a = \frac{11}{12} g$

Substitute into equation (2): $T = 5m (g - \frac{11}{12} g)$

$\Rightarrow T = \frac{55mg}{12}$

GURU TIP
If the inclined plane is rough there will be a frictional force opposing motion – see pages 76-77.

Forces in two dimensions

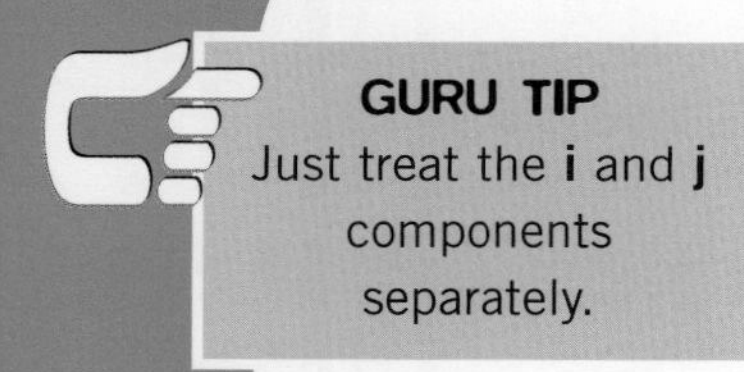

GURU TIP
Just treat the **i** and **j** components separately.

Newton's Second Law in two dimensions

Using Newton's Second Law in two dimensions involves vectors.

EXAMPLE A particle of mass 2kg is acted on by forces of ($10\mathbf{i} + 8\mathbf{j}$) N, ($24\mathbf{i} - 26\mathbf{j}$) N and ($-8\mathbf{i} - 6\mathbf{j}$) N. Find the magnitude of the acceleration.

SOLUTION 1. First, find the resultant force $\sum \mathbf{F}$. Remember to keep the **i** and **j** components separate.

$\sum \mathbf{F} = (10\mathbf{i} + 8\mathbf{j}) + (24\mathbf{i} - 26\mathbf{j}) + (-8\mathbf{i} - 6\mathbf{j}) = 26\mathbf{i} - 24\mathbf{j}$

2. Next, let the acceleration be $\mathbf{a}$ ms^{-2}, and use the equation $\sum \mathbf{F} = m\mathbf{a}$:

$26\mathbf{i} - 24\mathbf{j} = 2\mathbf{a} \Rightarrow \mathbf{a} = 13\mathbf{i} - 12\mathbf{j}$

3. Finally, use Pythagoras' Theorem to find the magnitude:

$|\mathbf{a}| = \sqrt{(12^2 + 13^2)} = \sqrt{313} = 17.7$

The magnitude of the acceleration is 17.7ms^{-2}.

Combining two perpendicular forces

Often you will not be given the vector in component form.

EXAMPLE A boat is being pulled into harbour by two tugs. Tug A is heading due north and tug B due east. The tensions in the light inextensible strings connecting the boat to A and B are 10000N and 20000N respectively.

(a) Find the angle at which the boat moves.

(b) Given that the resistance to the ship's motion due to the drag of the sea is 5000N and that the boat has a mass of 2000kg, find the acceleration of the boat.

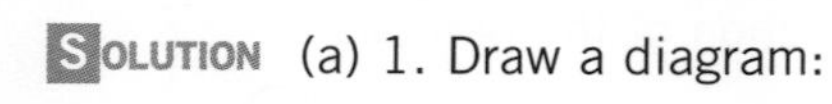

SOLUTION (a) 1. Draw a diagram:

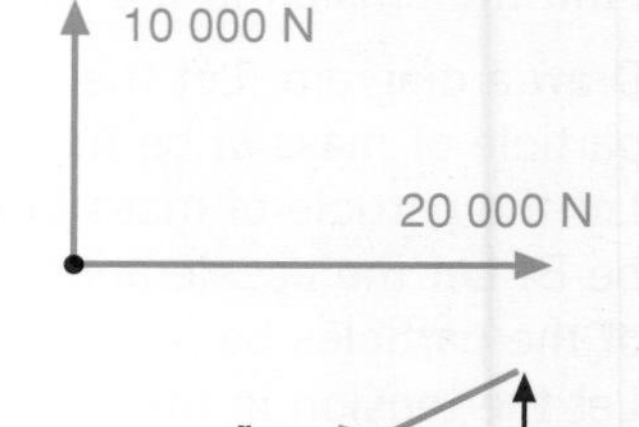

Add the forces 'nose to tail'. Let the resultant force be **r** newtons and let its angle from due east be θ.

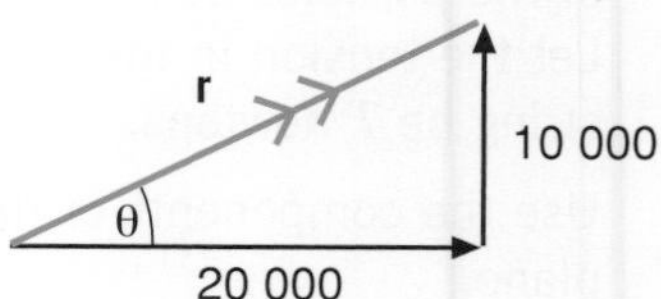

2. By trigonometry,

$\theta = \tan^{-1}\frac{10000}{20000} = 27°$ to the nearest degree.

(b) By Pythagoras' Theorem,

$|\mathbf{r}| = \sqrt{(10000^2 + 20000^2)} = 22360$

The problem is now one-dimensional. Take the direction of the boat as positive.

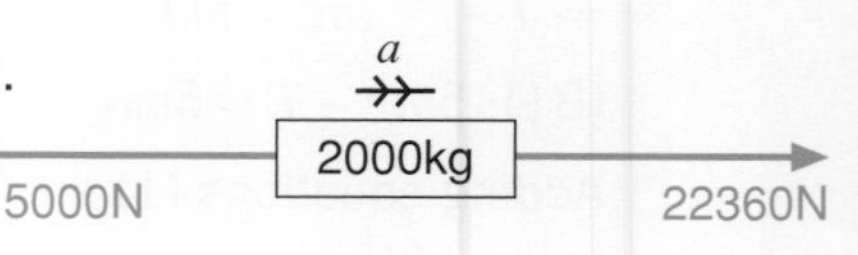

The equation of motion for the boat is:

$22360 - 5000 = 2000a \Rightarrow a = 8.68$

The acceleration of the boat is 8.68ms^{-2}.

Finding the resultant of any number of forces

To find the resultant of any number of forces, resolve them into two perpendicular components. See pages 56-57.

EXAMPLE Find the resultant of the following system of forces:

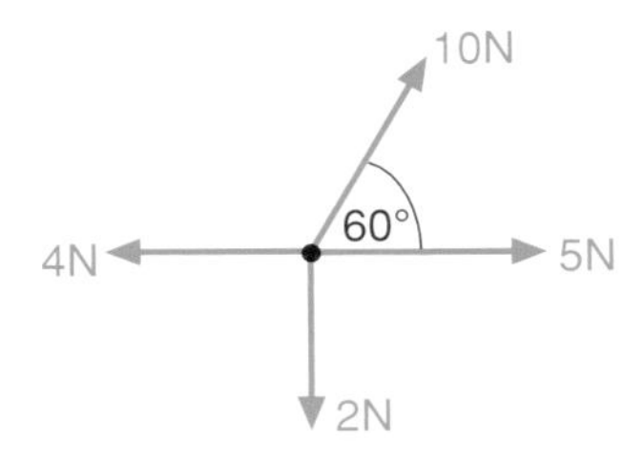

SOLUTION 1. First find the components of the 10N force:

Horizontal component $X = 10 \cos 60° = 5$

Vertical component $Y = 10 \sin 60° = 8.66$

2. Let the resultant be **R** newtons. The horizontal and vertical components of **R** are just the sums of the horizontal and vertical components of all the forces:

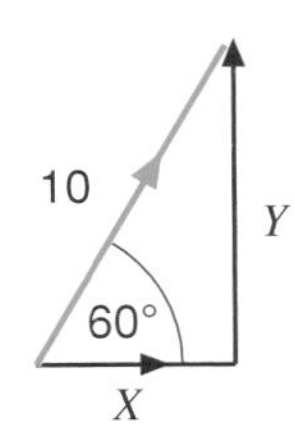

$(\rightarrow)$: $5 + 5 - 4 = 6$

$(\uparrow)$: $8.66 - 2 = 6.66$

3. So the magnitude of **R** is

$|\mathbf{a}| = \sqrt{(6^2 + 6.66^2)} = 9.0$

and the angle of **R** is

$\theta = \tan^{-1} \frac{6.66}{6} = 48°$

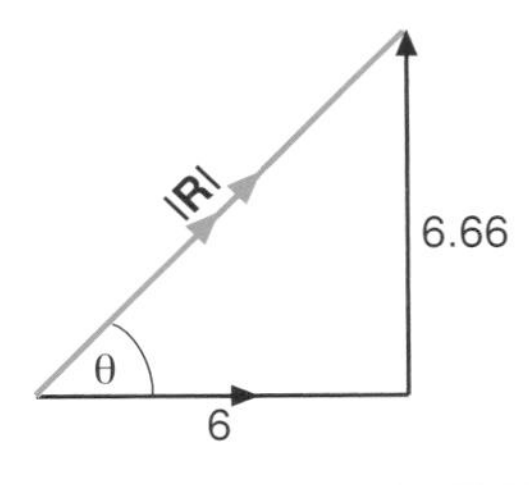

So the four forces have the same effect as one 9.0N force at an angle of 48° to the 5N force.

This could also be solved by scale drawing.

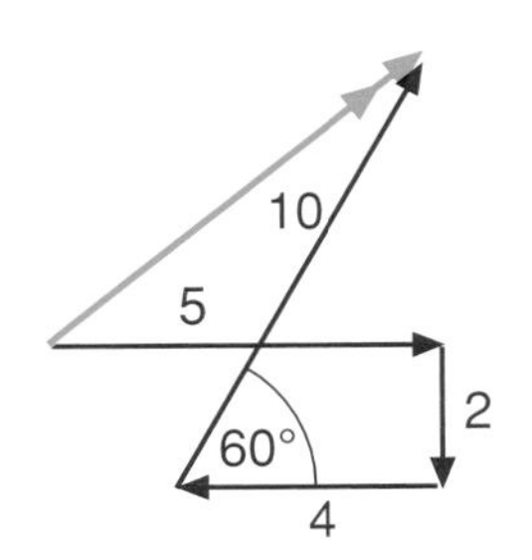

GURU TIP
Remember, vectors are added 'nose to tail'.

KEY SKILLS
Modelling a real-life situation using vectors and components in multi-stage calculations gives evidence to satisfy the criteria for Key Skills N3.2.

You could also use and compare alternative methods: graphical – scale drawing

numerical – components
This satisfies the criteria for N3.2 and N3.3.

Friction

Wherever there is a contact between two surfaces there is friction, unless the surfaces are smooth.

Friction is usually considered on an inclined plane.

Imagine a book on a table. What happens as one end of the table is slowly lifted?

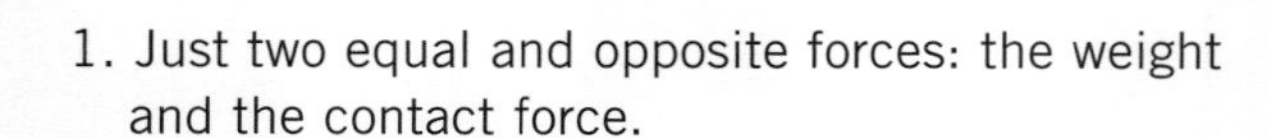

1. Just two equal and opposite forces: the weight and the contact force.

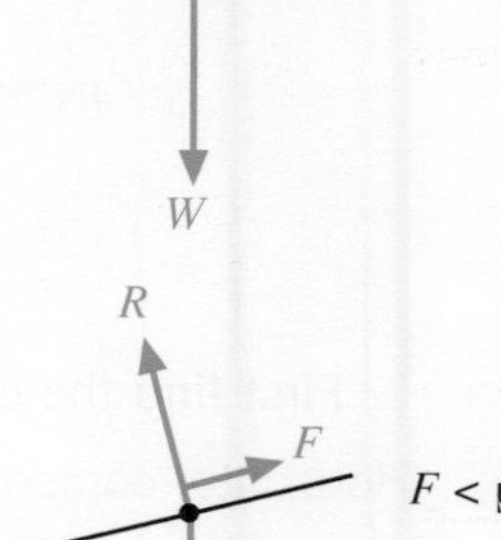

2. The frictional force opposes the component of the weight that is trying to move the book down the table. As the angle increases, so does the component of the weight along the table, and so does the frictional force. Until...

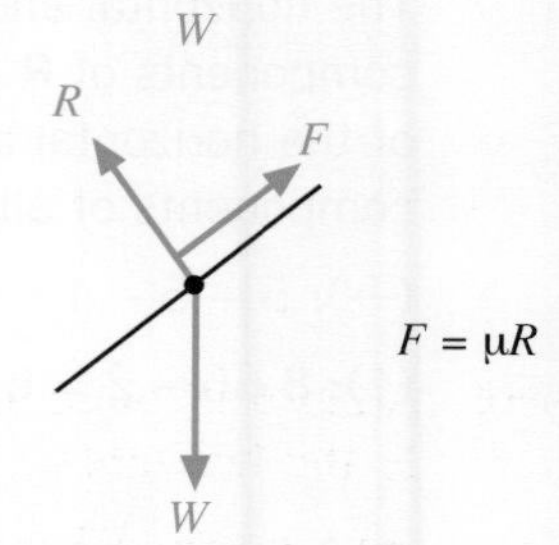

3. The component of the weight along the table is the same as the maximum frictional force. This is the point of **limiting equilibrium**.

If a body is in a state of limiting equilibrium, *either* it is still stationary, but on the point of moving, *or* it is moving at a constant velocity. The frictional force has reached its maximum value.

If the angle of inclination is increased further, the component of the weight along the table will be greater than the maximum frictional force. There will be a resultant force along the table, and hence acceleration.

The frictional force is *lazy*. It does as little as it has to. It always opposes motion.

Between the particle and the surface there is always a **coefficient of friction** μ.

> If F is the frictional force and R is the contact force, then
>
> $F \leq \mu R$
>
> If $F < \mu R$ then the particle is stationary. If $F = \mu R$ then the particle is in limiting equilibrium or moving.

Let's look at the inclined plane in more detail. What does 'component of the weight along the slope' mean?

When we resolve a force into two components, we usually choose the rightwards and upwards directions. But we are actually free to resolve them into any two directions. In questions involving an inclined plane, it is usually best to resolve along the line of the plane and at right angles to it. Then the only force whose components you will need to calculate is the weight.

So

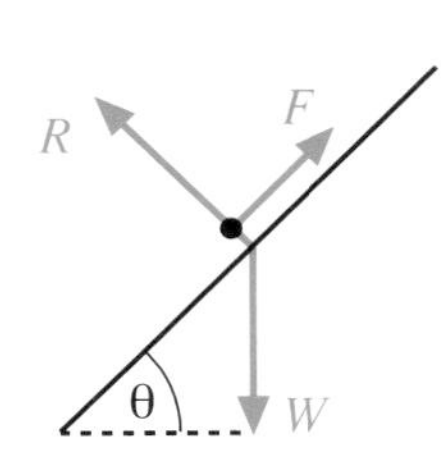

becomes

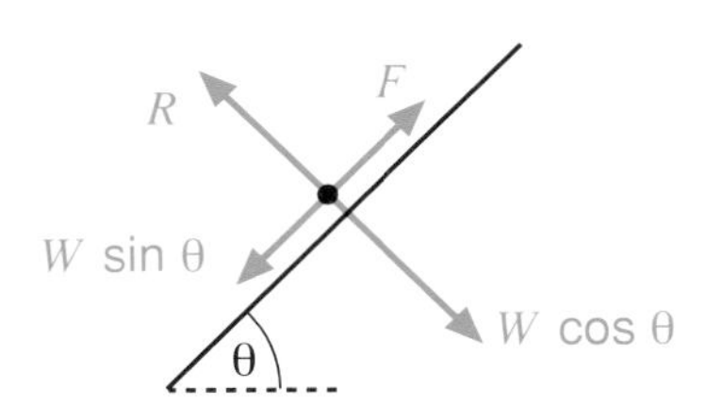

because

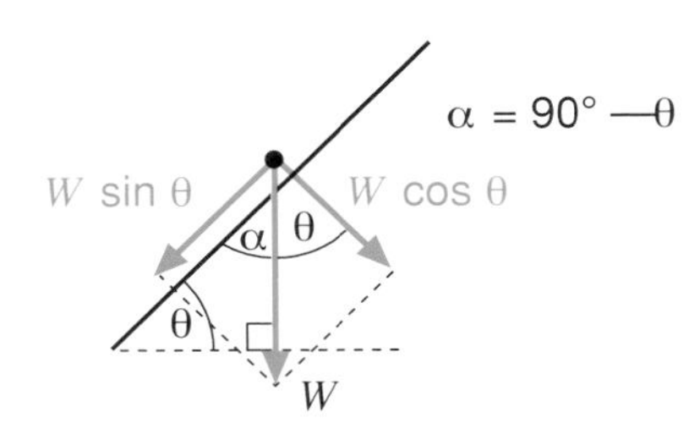

Mechanics

EXAMPLE A block of mass 2kg lies at rest on an plane inclined at an angle of 30° to the horizontal.

(a) Find the normal reaction of the plane on the block, R newtons, and the frictional force, F newtons.

(b) Given that the block is in limiting equilibrium, find the value of the coefficient of friction between the block and the plane.

GURU TIP
The component of the weight W down the slope is $W \sin \theta$ where θ is the inclination of the slope to the horizontal.

SOLUTION (a) 1. First draw a diagram.
Let W newtons be the weight of the block.

We know that $W = 2g$ and $\theta = 30°$.

2. Now resolve the forces. Take the downwards direction as positive.

Resolving perpendicular to the plane: $R - 2g \cos 30° = 0$
(the block stays on the plane)

$\Rightarrow R = 2 \times 9.8 \times 0.866 = 17.0$

Resolving parallel to the plane: $2g \sin 30° - F = 0$ (equilibrium)

$\Rightarrow F = 2 \times 9.8 \times 0.5 = 9.8$

The normal reaction is 17.0N and the frictional force is 9.8N.

(b) In limiting equilibrium, $F = \mu R \Rightarrow \mu = \frac{F}{R} = \frac{9.8}{17.0} = 0.58$

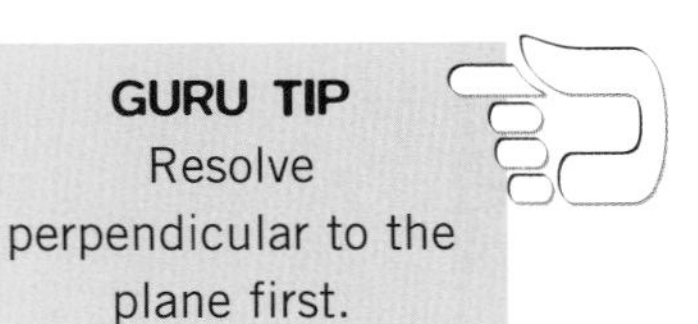

GURU TIP
Resolve perpendicular to the plane first.

Forces in equilibrium

Remember, Newton's First Law tells us that unless a resultant force acts on a body it will remain at rest, or if it is already moving, it will continue to move with a constant velocity.

A body that is at rest or travelling with constant velocity is said to be in **equilibrium.** Newton's First Law tells us that this happens when the resultant force is zero.

> For a body in equilibrium, $\sum \mathbf{F} = \mathbf{0}$

The components of the forces acting on the body in any direction must sum to zero.

The triangle of forces

Remember also the polygon of forces. This is the diagram you get if you draw the forces head to tail. If the resultant of a set of forces is zero, the polygon of forces forms a closed shape. For three forces, you get a 'triangle of forces'.

EXAMPLE A block of mass 2kg lies at rest on an plane inclined at an angle of 30° to the horizontal.

(a) Find the normal reaction of the plane on the block, R newtons, and the frictional force, F newtons.

(b) Given that the block is in limiting equilibrium, find the value of the coefficient of friction between the block and the plane.

SOLUTION Draw a diagram.

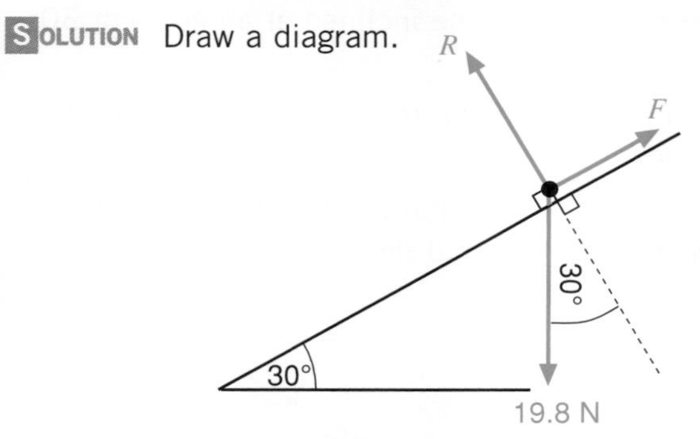

(a) Equilibrium ⇒ the triangle of forces is closed. Add the vectors nose to tail.

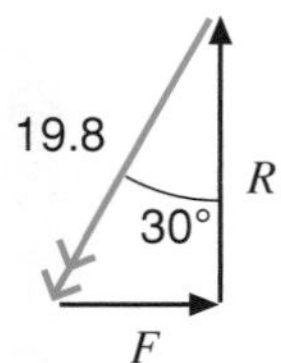

Now it's just a matter of trigonometry.

$F = 19.6 \sin 30° = 9.8$

$R = 19.6 \cos 30° = 17.1$

(b) $\mu = \frac{F}{R} = 0.58$

Another type of question involving three forces in equilibrium involves a body suspended from a ceiling supported by two strings.

EXAMPLE Coco the clown, who weighs 100kg, is to be suspended by two strings as shown. Find the tension in each string.

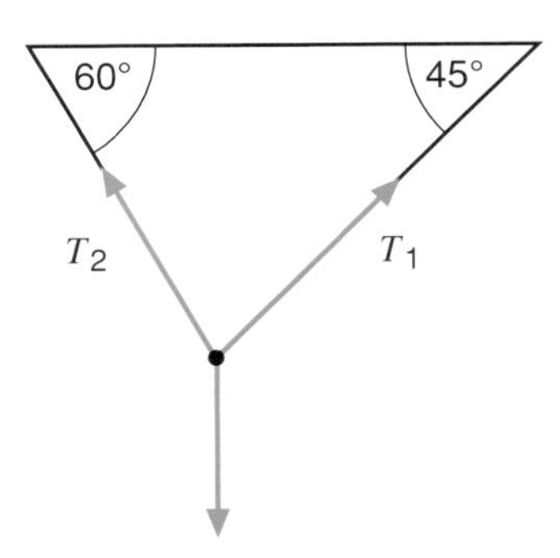

SOLUTION Copy the diagram and mark on the forces and any angles required.

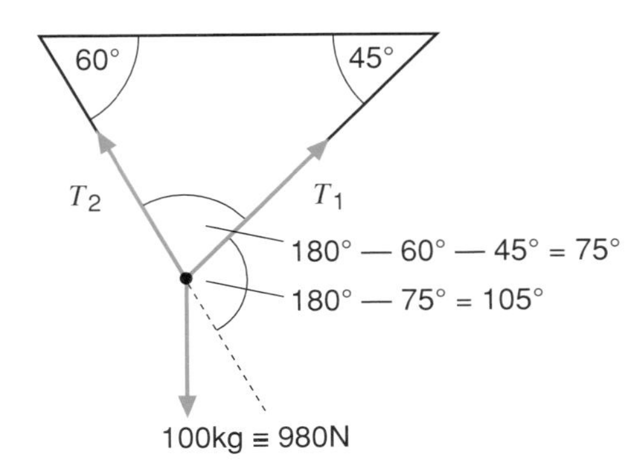

Then draw a triangle of forces.

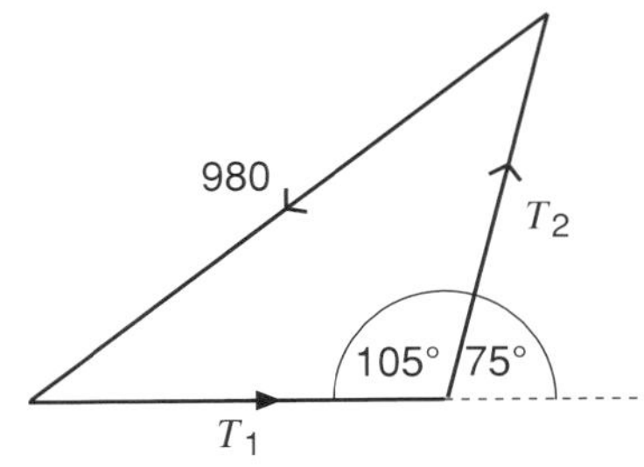

Only use scale drawing if the question says you can. Otherwise use the sine rule:

$$\frac{T_1}{\sin 30°} = \frac{T_2}{\sin 45°} = \frac{980}{\sin 105°}$$

So

$$T_1 = 980 \times \frac{\sin 30°}{\sin 105°} = 507$$

$$T_2 = 980 \times \frac{\sin 30°}{\sin 105°} = 717$$

The tensions in the two strings are 507N and 717N.

You can also do this question by resolving forces.

GURU TIP
As the resultant of the forces is zero, you end up at your starting point.

Mechanics

Moments

In Ancient Greece there lived a Guru called Archimedes for whom moments were a favourite topic. They may not be one of yours, but the idea is really quite simple.

When opening a big heavy door, you want to push it as far away from the hinge as possible. Why? Suppose the door is 4m wide and you can push with a force of 20N. Compare the effect of pushing the door in the middle and at the edge.

The turning effect depends on the size of the force and the distance of the force from the hinge.

4m
2m

Turning effect about the hinge = 20N × 2m = 40Nm

20N

Turning effect about the hinge = 20N × 4m = 80Nm

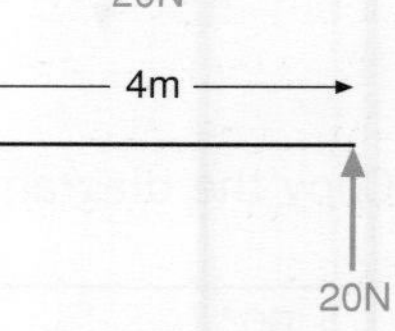

The turning effect is greater if you push the door at the edge. Another way of looking at this is that you need to work harder to open the door if you push at the middle than if you push at the edge.

These 'turning effects' are called **moments**. They are measured in Newton-metres (Nm).

In general:

> The moment of a force **F** about a point O is $|\mathbf{F}| \times d$
>
> where d is the (perpendicular) distance from O to (the line of action of) the force.

To describe a moment fully, you need to give the centre you are taking moments about and the size and direction (either clockwise or anticlockwise) of the force.

This diagram shows a see-saw:

The moment of force **A** about O is 10N × 3m = 30Nm (anticlockwise).

The moment of force **B** about O is 15N × 2m = 30Nm (clockwise).

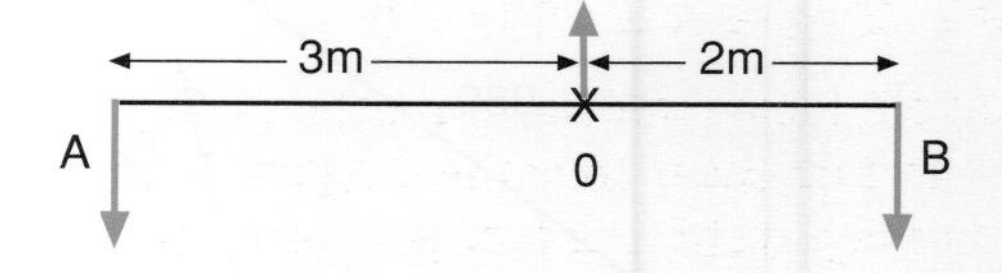

> For any body in equilibrium, the clockwise and anticlockwise moments are equal.

As the clockwise and anticlockwise moments are equal, the see-saw is in equilibrium. Notice the contact force at the pivot. How big is it?

Most exam questions on this topic are about conditions for equilibrium for a plank. The plank will either be resting on supports (in which case you will have contact forces) or suspended (in which case you will have tensions in the supporting strings). All these questions are tackled in the following way:

- Step 1: Draw a diagram.
- Step 2: Take moments about any point. The clockwise and anticlockwise moments must be equal.
- Step 3: Resolve vertically ($\sum \mathbf{F} = \mathbf{0}$).

The planks are modelled as uniform rods. This means that their weight acts at the centre. So just draw a line to represent the plank and draw an arrow coming down from the middle to represent its weight. Of course, if the plank is modelled as a light rod, you will not need this arrow.

GURU TIP
Remember to include the contact force R on your diagram.

EXAMPLE Fred, Ginger and Rover the dog are sitting on a uniform plank AB of length 8m and mass 20kg, pivoted 3m from A. Fred weighs 80kg and Rover weighs 10kg. Ginger says she weighs 40kg.

When Fred sits at A, Ginger sits at B and Rover sits 3m from Ginger, the plank is in equilibrium. Is Ginger telling the truth?

SOLUTION Draw a diagram. Let the mass of Ginger be M kg.

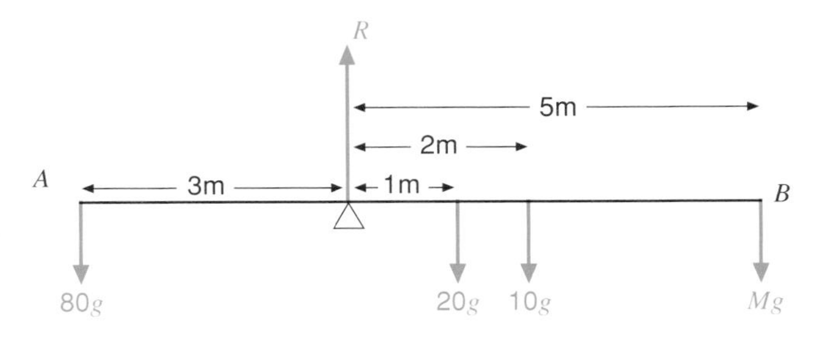

Take moments about the pivot. The clockwise and anticlockwise moments must be equal.

$80g \times 3 = 20g \times 1 + 10g \times 2 + Mg \times 5$

(The g factors cancel, as they often do in these problems.)

$240 = 20 + 20 + 5M$

$200 = 5M \Rightarrow M = 40$

Ginger does indeed weigh 40kg. (Did you doubt her?)

You can take moments about any point. In the above example, it was easiest to take them about the pivot, otherwise we would have had to calculate R. This is why the distances marked on the diagram were from this point. So it's best to choose what point to take moments about before marking the distances on your diagram.

If you want to find R, just resolve all the forces vertically:

$\uparrow: R - 80g - 20g - 10g - Mg = 0$

$M = 40\text{kg} \Rightarrow R = 150g = \text{N}$

If you are lucky, the exam question will give you the diagram rather than a long story. In this case, copy the diagram from the exam paper, marking in all forces and any missing distances that you need. Then take moments and resolve. Let's look at an example involving a suspended rod.

EXAMPLE The diagram shows a uniform rod AB of length 3m and mass 10kg. It is supported by two strings attached at A and C. Find the tensions in the two strings.

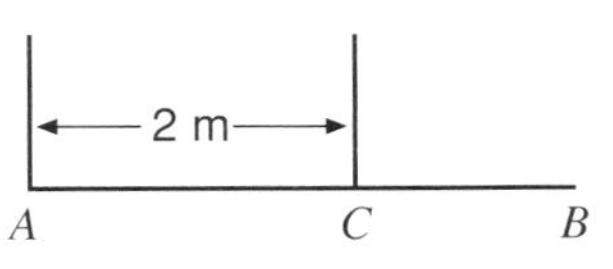

SOLUTION Copy the diagram and draw all the forces on it. Let the tensions in the strings at A and C be T and S newtons respectively.

Take moments about A:

$10g \times 1.5 = S \times 2 \Rightarrow 147 = 2S$
$\Rightarrow S = 73.5\text{N}$

Resolve vertically:

$\uparrow: T + S = 10g \Rightarrow T + 73.5 = 98$
$\Rightarrow T = 24.5\text{N}$

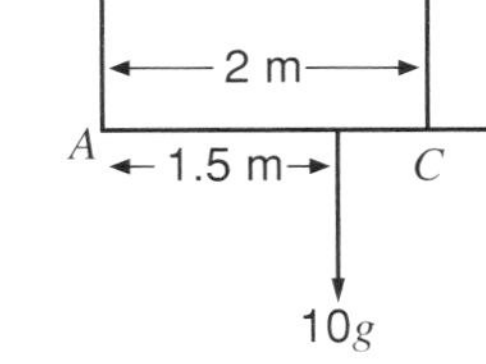

Mechanics

Centre of mass

The centre of mass of an object is the point at which it balances. For example, if you spin a plate on your finger, it is best to put your finger directly under the centre of mass.

GURU TIP
Make as much use as you can of symmetry in these questions.

Finding the centre of mass by symmetry

Model the plate as a uniform circular lamina. (Remember: 'uniform' means the density is constant; 'lamina' means it has no thickness.)

By symmetry, the centre of mass must be in the centre.

Similarly for a rectangle $OABC$, 6 units by 4 units:

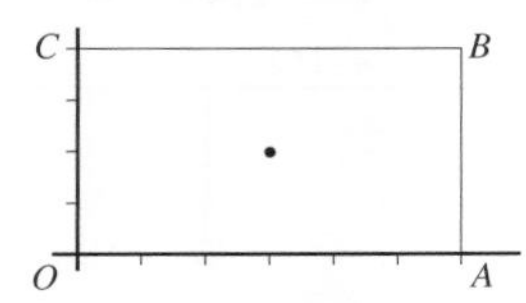

By symmetry, the centre of mass is at (3,2).

Finding the centre of mass of a system of particles

EXAMPLE Given a light rod of length 10m with particles A, B and C of mass 2kg, 4kg and 10kg respectively attached as shown. Find the position of the centre of mass.

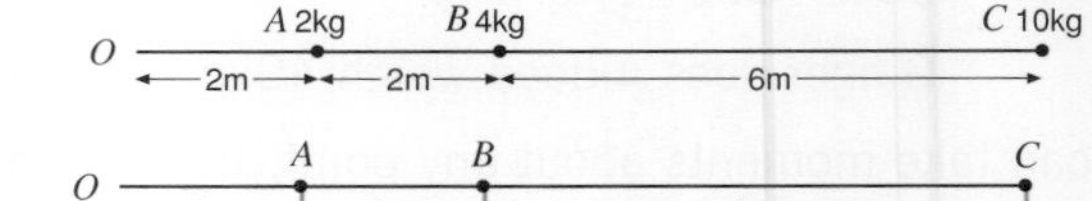

SOLUTION Draw in all the forces.

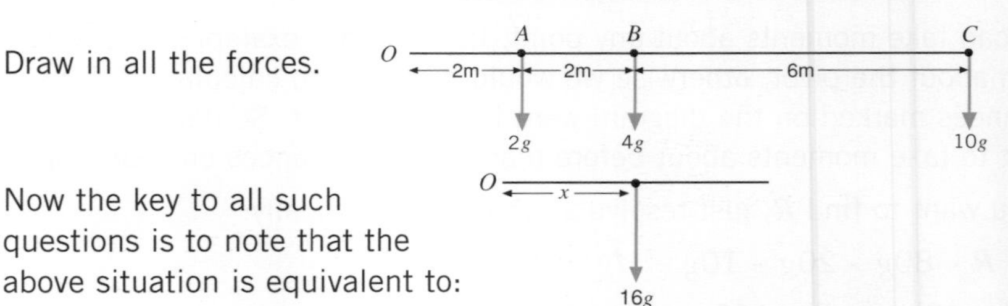

Now the key to all such questions is to note that the above situation is equivalent to:

where x metres is the distance from O of the centre of mass. It is as if the total mass were acting through the centre of mass.

So taking moments about O:

$2g \times 2 + 4g \times 4 + 10g \times 6 = 16g \times x$

Cancelling g:

$2 \times 2 + 4 \times 4 + 10 \times 6 = 16x$

$80 = 16x$

$x = 5$

The centre of mass of the system is 5m from the left-hand end of the rod.

A good way to tackle such questions is to use a table:

	A	B	C	Total
mass	2	4	10	16
distance from O	2	4	6	x
'moment'	4	16	60	$80 = 16x$

(Notice that our 'moments' are not strictly moments, as a factor of g has been cancelled.)

EXAMPLE An earring is formed by cutting a circle of radius 2cm from a uniform circular lamina of radius 4cm as shown. Find the position of the centre of mass of the earring.

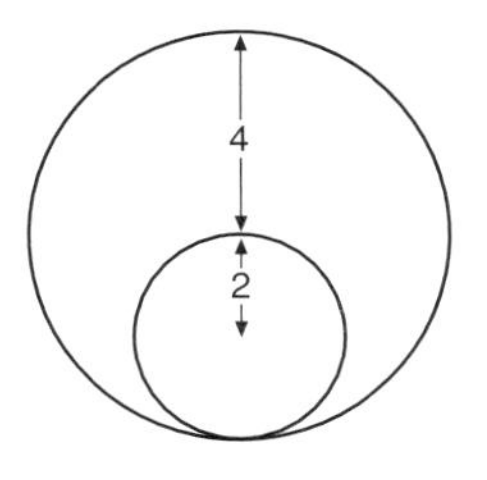

GURU TIP Draw the axis you are using on your diagram.

SOLUTION As the earring has a line of symmetry, the centre of mass will lie on this line. Take this line as the y axis and let $\bar{y}$ be the distance up it of the centre of mass.

Draw a table. As the lamina has a uniform density, masses are proportional to areas. For convenience, in the table we divide the areas by 4π to obtain the relative masses.

	big circle	little circle	difference (earring)
area	16π	4π	12π
relative mass	4	1	3
y coordinate	4	2	$\bar{y}$
'moment'	16	2	$16 - 2 = 3\bar{y}$

$16 - 2 = 3\bar{y} \Rightarrow 3\bar{y} = 14 \Rightarrow \bar{y} = \frac{14}{3}$

So the centre of mass of the earring is 4.67cm up the axis of symmetry.

For a two-dimensional shape without an axis of symmetry, you have to consider the x and y coordinates of the centre of mass separately.

EXAMPLE Find the centre of mass of the uniform lamina shown.

SOLUTION Take B as the origin, BC as the x axis and BA as the y axis. Let the centre of mass be $(\bar{x},\bar{y})$.

Taking moments in both coordinates:

	tall rectangle	square	total (lamina)
area	12	4	16
relative mass	3	1	4
y coordinate	3	1	$\bar{y}$
'y moment'	9	1	$9 + 1 = 4\bar{y}$
x coordinate	1	3	$\bar{x}$
'x moment'	3	3	$3 + 3 = 4\bar{x}$

So

$\bar{x} = 1.5$

$\bar{y} = 2.5$

The centre of mass of the lamina is at a point 1.5cm from AB and 2.5cm from BC.

Angle of inclination

If a lamina is freely suspended and in equilibrium, its centre of mass will lie vertically under the point of suspension.

GURU TIP You can find the angle of a suspended lamina by drawing a diagram with the centre of mass underneath the point of suspension.

Impulse and momentum

'Impulse' is used when a large force acts for a short time: for example, when two balls collide in a game of pool, or in a 'Newton's cradle' (your teacher may have one of these locked away somewhere), or a bat hitting a ball. The equation you need is:

$$\mathbf{I} = \mathbf{F}t = m\mathbf{v} - m\mathbf{u}$$

(This is derived from $\mathbf{F} = m\mathbf{a}$ and $\mathbf{v} = \mathbf{u} + \mathbf{a}t$. Can you see how?)

The product of the mass and the velocity is called the **momentum**. So we can also write:

impulse = final momentum – initial momentum

impulse = change in momentum

The units of both momentum and impulse are Newton-seconds (Ns).

EXAMPLE A fielder catches a cricket ball of mass 0.2kg travelling at 20ms^{-1}. What impulse acts on her hands?

SOLUTION Resolving in the direction of travel: $m = 0.2$kg, $v = 0$ (the ball is brought to rest in being caught), $u = 20\text{ms}^{-1}$.

$I = mv - mu = 0.2 \times 0 - 0.2 \times 20$

$= -4$ Ns

(The negative sign shows that the impulse is in the opposite direction to the velocity.)

Questions on impulse tend to be hidden away in questions on momentum.

GURU TIP
Impulse and momentum are vector quantities, so be careful with signs.

The Principle of Conservation of Momentum

This grand title is often abbreviated to PCM. (COM is also sometimes used.) It tells us that when bodies collide, momentum is conserved. That is:

total momentum before = total momentum after

$$m_1\mathbf{u}_1 + m_2\mathbf{u}_2 = m_1\mathbf{v}_1 + m_2\mathbf{v}_2$$

$\mathbf{u}_1$ $\mathbf{u}_2$! $\mathbf{v}_1$ $\mathbf{v}_2$

m_1 m_2 m_1 m_2

$m_1\mathbf{u}_1 + m_2\mathbf{u}_2 = m_1\mathbf{v}_1 + m_2\mathbf{v}_2$

That's all there is to it. Whatever the question, start by drawing this diagram, with the quantities given, and writing out the PCM equation.

A special case is when bodies couple or coalesce on colliding.

EXAMPLE A railway truck of mass 3000kg travelling at 4 ms^{-1} couples with a stationary truck of mass 1000kg. Find:

(a) the speed of the trucks after they have coupled

(b) the impulse exerted by the lighter truck on the heavier one

SOLUTION (a) Draw a diagram:

4 0 ! v

3000 1000 3000—1000

PCM($\rightarrow$): $3000 \times 4 + 1000 \times 0 = (3000 + 1000) \times v$

Statistics – introduction

Statistics is concerned with the collection, representation and analysis of data. Probability models are used to describe situations where there is an element of chance or randomness.

Governments and businesses increasingly use statistical analysis and probability models to help them plan for the future. An understanding of statistical techniques is essential in many subjects.

Your calculator should be able to perform many of the statistical analyses that you will need. Make sure that you know how to use it. But your calculator can't do everything for you. It's important that you understand the techniques.

Glossary of terms

A **variable** is a property of something, which can take on different values.

Data are measurements (or observations) of a variable.

Variables may be either **qualitative** or **quantitative**. A qualitative variable tells you some quality of something. A quantitative variable tells you some numerical quantity associated with something.

Quantitative variables are either **discrete** or **continuous**. Discrete data can only take certain values. Continuous data can take any value (in a given range). For example:

Variable	Data	Qualitative or quantitative?	Discrete or continuous?
favourite soap	EE, BS, CS	qualitative	—
time breath held	89.9s, 10.1s, 54.0s	quantitative	continuous
money in pocket	£1.29, £20.32, 21p	quantitative	discrete

Primary data are data that you collect yourself. **Secondary** data are data obtained from other sources, such as books, the Internet and CD ROMs.

KEY SKILLS

In studying Statistics at AS Level, you will have many opportunities to develop the Key Skills for Application of number at Level 3. There are also opportunities to demonstrate Information Technology skills by using spreadsheets or other software as a modelling tool (Key Skill IT3.2 – see chart on page 6).

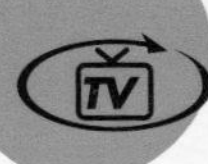
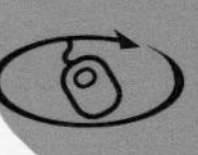

Data collection and probability

Suppose you were asked the question: 'How tall are Year 12 students in your school or college?' If you just measured the first person you saw, your data would not be representative. On the other hand, the population is too large to measure everybody, and even if you could you would need to summarise the data somehow rather than present a long list of figures.

What you would actually do is measure a few people (**survey a sample**) out of the whole year group (the **population**) and calculate numerical measures (**statistics**) to summarise the data. The sample mean $\bar{x}$, for example, gives an **estimate** of the population mean (labelled by the Greek letter μ). The population mean is a **parameter** of the population. *Statistics estimate parameters.*

When collecting and analysing data, you want to find numerical measures (statistics) that give a good estimate of the value of the parameter you are interested in.

WEB TIP
There's more about data on the AS Guru™ website.

Sampling

When you take a sample, you must make two key decisions:

- What size sample to take. A bigger sample gives better results but costs more to collect.
- How to choose your sample. You want your sample to be representative, that is, to reflect the whole population in its make-up.

The main sampling methods are:

- **Random sampling**. Each member of the population has the same chance of being chosen. Random-number tables or generators (your calculator may have one of these) are useful in choosing the sample. The problem with random sampling is that it does not necessarily ensure that your sample is representative.
- **Stratified sampling**. This ensures that the sample is representative. The population is divided into categories, and the sample is chosen so that the numbers of members chosen from the different categories are in the same proportions as in the whole population. For example, if you were researching people's voting intentions, you may want to make sure that your sample has the same proportion of people from each social class as the whole population. Once you have decided how many to choose from each category, use random sampling within each category.
- **Systematic sampling**. Items are chosen in a regular manner. For example, you may ask every tenth person in a queue, or test every fifth item coming off a production line.
- **Purposive sampling**. This is used when only part of the population is of interest. For example, if you were interested in the amount of money people spent on dog food, you would only want to survey those members of the population that owned a dog.

The language of probability

The **sample space** is the set of all possible outcomes of an **experiment**.

For example, if you roll a die you are performing an experiment. The sample space is the set {1,2,3,4,5,6}.

An **event** is a group of possible outcomes.

When tossing a coin, some examples of events are:

- the number on the die is even (A)
- the number on the die is prime (B)
- the number on the die is not even (A′)

Events are usually labelled by capital letters.

The two events A and A′ are **complementary**. That is, between them they account for every possible outcome in the sample space. A′ is the event 'not A'.

If P(X) stands for the probability of an outcome X, then

$$P(A) + P(A') = 1$$
$$P(A') = 1 - P(A)$$

Calculating the probability of an event

If all outcomes are equally likely:

$$\text{(probability of an event)} = \frac{\text{(number of outcomes in the event)}}{\text{(total number of outcomes)}}$$

In the example above, the event A ('the number on the die is even') contains 3 outcomes (2, 4 and 6), so $P(A) = \frac{3}{6} = \frac{1}{2}$. The event B ('the number on the die is prime') also contains 3 outcomes (2, 3 and 5), so $P(B) = \frac{3}{6} = \frac{1}{2}$.

What is the probability that the number is both even and prime?

This is written: $P(A \cap B)$ ('the probability of A and B').

The number of even prime numbers in the sample space is just 1. So $P(A \cap B) = \frac{1}{6}$.

- $A \cap B$ means 'A and B'.
- $A \cup B$ means 'A or B'.

As ever, a diagram helps to clarify matters. **Venn diagrams** are a convenient way of displaying information in probability problems.

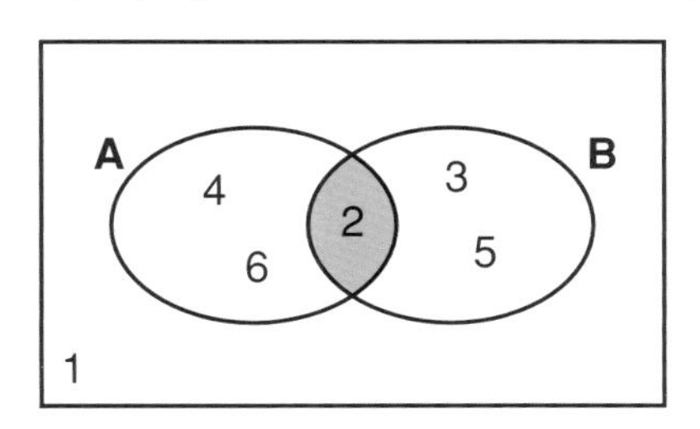

WEB TIP
You can practise the use of this notation on the AS Guru™ website.

Numerical methods of summarising data

The two types of summary statistic

It is often useful to represent the whole population by two numbers: a **measure of location** (where the data are) and a **measure of dispersion** (how spread out the data are).

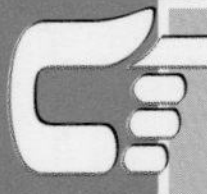

GURU TIP
When calculating the mode, remember that it may not be unique: for example, {1,2,2,3,4,4,5} has two modes, 2 and 4.

Measures of location

The three most common measures of location are:

- **mode**: the most frequent value of the variable ('most')
- **median**: the middle value of the ordered data set ('middle')
- **mean**: the sum of the values divided by the number of values ('fair shares')

The mode and median are often easier to calculate than the mean, and are not affected by extreme values. However, they do not have such nice mathematical properties, and the mean is the best representative of the whole data set.

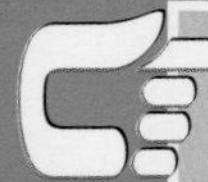

GURU TIP
When calculating the median, if there is an even number of observations take the mean of the middle two.

If there are n observations arranged in order then the position of the median is $\frac{1}{2}(n+1)$. If this is not an integer, take the mean of the numbers on either side.

EXAMPLE Find the median of these observations: 1, 1, 2, 2, 3, 3, 4, 4, 5, 5, 7, 8, 9.

SOLUTION There are 13 numbers in the list. $\frac{1}{2}(13+1) = 7$. So the median is the 7th number in the list, which is 4.

EXAMPLE Find the median of these observations: 1, 1, 2, 2, 3, 3, 4, 4, 5, 5, 7, 8.

SOLUTION There are 12 numbers in the list. $\frac{1}{2}(12+1) = 6\frac{1}{2}$. So the median is the mean of the 6th and 7th numbers, which is $\frac{1}{2}(3+4) = 3\frac{1}{2}$.

Notice that in this case the median does not appear in the list.

Measures of dispersion

In addition to the median, the lower and upper **quartiles** are useful statistics. The **lower quartile** is one-quarter of the way along the ordered data set, and the **upper quartile** is three-quarters of the way along it.

The four most common measures of disperion are:

- **range**: the difference between the greatest and least values
- **interquartile range**: the difference between the upper and lower quartiles
- **variance**: the mean of the squared deviations from the mean
- **standard deviation**: the positive square root of the variance

Mean and variance

The most important measures of location and dispersion are the mean and variance respectively.

WEB TIP
There's more about variance on the AS Guru™ website.

For raw data, the mean is $\bar{x} = \frac{\Sigma x}{n}$

and the variance is $\frac{\Sigma (x - \bar{x})^2}{n} = \frac{\Sigma x^2}{n} - \bar{x}^2$

EXAMPLE The following data refer to the number of hours of sunshine in Sunnyville in June for the last 12 years. Find the mean and variance.

191, 177, 203, 204, 208, 183, 191, 184, 140, 137, 182, 196

SOLUTION Mean: $\bar{x} = \frac{\sum x}{n} = \frac{2196}{12} = 183$ hours

To find the variance, calculate the deviations from the mean:

data item	191	177	203	204	208	183	191	184	140	137	182	196
deviation from mean	+8	–6	+20	+21	+25	0	+8	+1	–43	–46	–1	+13
squared deviation	64	36	400	441	625	0	64	1	1849	2116	1	169

Mean of squared deviations from the mean = $\frac{5766}{12} = 480.5$

Variance = 480.5 hours^2

A note about units. The squaring of the deviations leads to the squaring of the units for the variable – in the above example, the number of hours of sunshine. The standard deviation, however, being the square root of the variance, has the same units as the data.

(standard deviation) = √(variance)

In the above example, the standard deviation is 21.9 hours.

You may also be asked to calculate the mean and variance from summary statistics.

EXAMPLE The marks, x, obtained by 40 students in their S1 exam were recorded. The following results were obtained.

$\sum x = 2840$

$\sum x^2 = 206480$

Calculate the mean and standard deviation.

SOLUTION Mean = $\frac{\sum x}{n} = \frac{2840}{40} = 71$ marks

Variance = $\frac{\sum x^2}{n} - \bar{x}^2 = \frac{206480}{40} - 71^2 = 121$

Standard deviation = √(variance) = 11 marks

KEY SKILLS
Finding the mean and variance of a data set of 50 items or more meets the criteria for Key Skills N3.1. If the data was downloaded from the Internet, you would be demonstrating use of Information Technology.

GURU TIP
You will not need to work through questions like this laboriously. Your calculator can do all this for you. Make sure you learn how to use it for this.

Calculating the mean and variance

Calculating the mean from a frequency table

For a frequency distribution f, the mean is $\bar{x} = \frac{\Sigma xf}{\Sigma f}$

and the variance is $\frac{\Sigma (x - \bar{x})^2 f}{\Sigma f} = \frac{\Sigma x^2 f}{\Sigma f} - \bar{x}^2$

EXAMPLE Given the following data, where the variable x is the number of eggs laid in a given week on a farm with 100 chickens. Calculate the mean number of eggs.

x	$f(x)$
0	7
1	10
2	15
3	25
4	30
5	13

SOLUTION Add an extra column to the table, and calculate xf for each row. Write in the totals $\sum f(x)$ and $\sum xf(x)$.

x	$f(x)$	$xf(x)$
0	7	$0 \times 7 = 0$
1	10	$1 \times 10 = 10$
2	15	$2 \times 15 = 30$
3	25	$3 \times 25 = 75$
4	30	$4 \times 30 = 120$
5	13	$5 \times 13 = 65$
	$\sum f(x) = 100$	$\sum xf(x) = 300$

So the mean is $\bar{x} = \frac{300}{100} = 3$ eggs.

WEB TIP
There's more about variance on the AS Guru™ website.

Calculating the variance from a frequency table

Continuing with the above example, if we want to find the variance we just add another column to the table:

x	$f(x)$	$xf(x)$	$x^2f(x)$
0	7	$0 \times 7 = 0$	$0^2 \times 7 = 0$
1	10	$1 \times 10 = 10$	$1^2 \times 10 = 10$
2	15	$2 \times 15 = 30$	$2^2 \times 15 = 60$
3	25	$3 \times 25 = 75$	$3^2 \times 25 = 225$
4	30	$4 \times 30 = 120$	$4^2 \times 30 = 480$
5	13	$5 \times 13 = 65$	$5^2 \times 13 = 325$
	$\sum f(x) = 100$	$\sum xf(x) = 300$	$x^2f(x) = 1100$

The variance is $\sigma^2 = \frac{1100}{100} - 3^2 = 11 - 9 = 2$, and the standard deviation is $\sqrt{2} = 1.41$.

GURU TIP
Your calculator can work both these results out for you. You just have to enter the values and their frequencies. Make sure you know how to do this.

Estimating the mean and variance from grouped data

When individual values are not known, the midpoint of each group is used. The calculations of mean and variance are only estimates, since they do not use the exact data.

EXAMPLE The following grouped frequency table shows the number of goals scored by a group of 100 footballers. Estimate the mean and variance.

Goals	f
0–4	12
5–9	18
10–14	25
15–19	28
20–24	12
25–29	5

SOLUTION Extend the table:

Goals	f	Mid-point x	xf	x^2f
0–4	12	2	24	48
5–9	18	7	126	882
10–14	25	12	300	3600
15–19	28	17	476	8092
20–24	12	22	264	5808
25–29	5	27	135	3645
	$\sum f = 100$		$\sum xf = 1325$	$\sum x^2f = 22075$

The mean is $\bar{x} = \frac{\sum xf}{\sum f} = \frac{1325}{100}$ 1325/100 = 13.25.

The variance is $\frac{\sum x^2 f}{\sum f} - \bar{x}^2 = \frac{22075}{100} - 13.25^2 = 45.1875$.

The standard deviation = $\sqrt{45.1875} = 6.7$ goals.

KEY SKILLS
If using the method shown on the left for a piece of work, or a report, consider the limitations it imposes on the results.

GURU TIP
The midpoint is our best estimate for each of the values.

KEY SKILLS
To what degree of accuracy would it be sensible to give these results, given that they come from grouped data?

Statistics

Coding data

The calculations can often be greatly simplified by transforming the data in some way. This is called 'coding'.

EXAMPLE Find the mean and variance of the following set of marks:

100, 100, 104, 104, 108, 112, 116, 124, 124, 128

SOLUTION 1. Transform the data.

First take away the lowest value (100) from every value. This gives:
0, 0, 4, 4, 8, 12, 16, 24, 24, 28

2. Next divide by any common factor (in this case 4). This gives:
0, 0, 1, 1, 2, 3, 4, 6, 6, 7

Find the mean and variance of the resulting data set.

Mean = (0 + 0 + 1 + 1 + 2 + 3 + 4 + 6 + 6 + 7) ÷ 10 = 30 ÷ 10 = 3
Variance = $((0-3)^2 + (0-3)^2 + (1-3)^2 + \ldots (7-3)^2) \div 10$
= (9 + 9 + 4 + 4 + 1 + 0 + 1 + 9 + 9 + 16) ÷ 10 = 62 ÷ 10 = 6.2

To find the mean of the original data, undo the transformation:

1. Multiply by the common factor: 3 × 4 = 12
2. Add the lowest value: 12 + 100 = 112

To find the variance of the original data, just multiply by the square of the common factor: $6.2 \times 4^2 = 6.2 \times 16 = 99.2$

So for the original data, the mean is 112 and the variance is 99.2.

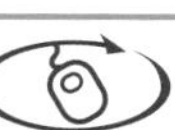

Graphical methods of summarising data

Stem-and-leaf diagrams

This is a simple but informative way of presenting data.

EXAMPLE Use a stem-and-leaf diagram to present the following data (ages of people at a party).

10, 22, 12, 34, 43, 32, 21, 35, 16, 28, 19, 38, 25, 21, 41, 37, 28, 15, 27, 18

SOLUTION 1. First make a 'tally chart': the numbers at the beginning of the rows (the 'stems') represent the tens digits; and you just write down the second digit (the 'leaf') of each number on the appropriate row:

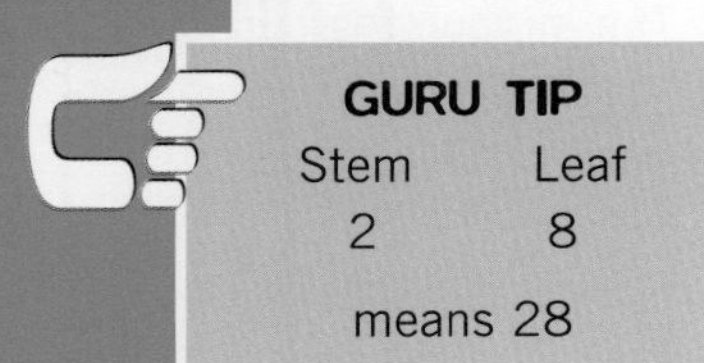

GURU TIP
Stem Leaf
2 8
means 28

1	0, 2, 6, 9, 5, 8
2	2, 1, 8, 5, 1, 8, 7
3	4, 2, 5, 8, 7
4	3, 1

2. Now you have to draw another diagram, putting the 'leaves' in order. At the right-hand end of each row put the frequency in brackets:

GURU TIP
Don't forget to put the 'leaves' in order.

1	0, 2, 5, 6, 8, 9	(6)
2	1, 1, 2, 5, 7, 8, 8	(7)
3	2, 4, 5, 7, 8	(5)
4	1, 3	(2)

To compare two sets of data you can use two stem-and-leaf diagrams 'back to back'.

EXAMPLE Draw stem-and-leaf diagrams to compare the following classes' exam results:

Teacher A: 31, 32, 38, 39, 40, 41, 42, 42, 44, 46, 46, 48, 49, 50, 54, 56, 57, 60, 61, 67

Teacher B: 37, 41, 44, 46, 49, 51, 54, 54, 55, 55, 56, 57, 57, 58, 59, 59, 62, 62, 64, 65

SOLUTION

Teacher A				Teacher B
(2)	2 1	3		(0)
(2)	9 8	3	7	(1)
(5)	4 2 2 1 0	4	1 4	(2)
(4)	9 8 6 6	4	6 9	(2)
(2)	4 0	5	1 4 4	(3)
(2)	7 6	5	5 5 6 7 7 8 9 9	(8)
(2)	1 0	6	2 2 4	(3)
(1)	7	6	5	(1)

Histograms

Histograms are the 'big brothers' of bar charts. A histogram is used to represent a continuous variable which has been summarised in a grouped frequency table. As the variable is continuous, there are no gaps between bars in a histogram.

On a histogram, the area of each bar is proportional to the frequency. So

(class width) $\times$ (frequency density) = $\kappa \times$ (frequency)

To make life easy, choose $\kappa = 1$. So (class width) × (frequency density) = (frequency). Rearranging:

$$(\text{frequency density}) = \frac{(\text{frequency})}{(\text{class width})}$$

EXAMPLE Complete the following grouped frequency table and draw a histogram, showing the heights of 60 men.

height x cm	Frequency f	Frequency density
$120 \le x < 130$	4	
$130 \le x < 140$	9	
$140 \le x < 150$	13	
$150 \le x < 160$	20	$20 \div 10 = 2.0$
$160 \le x < 170$	11	
$170 \le x < 190$	3	

SOLUTION Divide each frequency by its class range to get the frequency density.

height x cm	Frequency f	Frequency density
$120 \le x < 130$	4	$3 \div 10 = 0.3$
$130 \le x < 140$	9	$9 \div 10 = 0.9$
$140 \le x < 150$	13	$13 \div 10 = 1.3$
$150 \le x < 160$	20	$20 \div 10 = 2.0$
$160 \le x < 170$	11	$11 \div 10 = 1.1$
$170 \le x < 190$	3	$4 \div 20 = 0.2$

N3.3

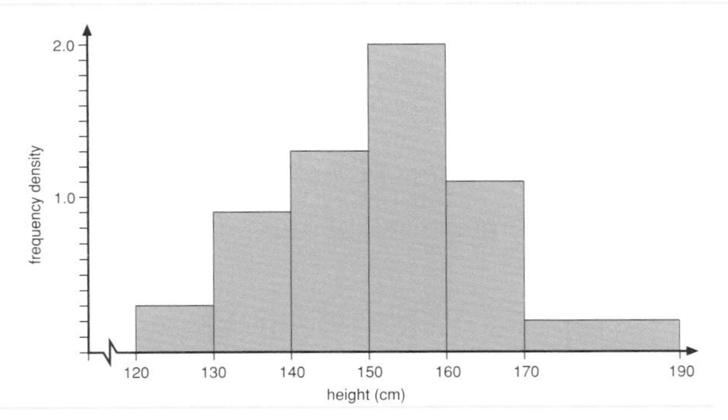

KEY SKILLS
Appropriate use of charts and diagrams, and the interpretation of results, meet the criteria fror Key Skills N3.3.

Relative frequencies

Relative frequency is useful when comparing two data sets. It gives the fraction of the total accounted for by each group:

$$\text{relative frequency} = \frac{\text{frequency}}{\text{total freqency}}$$

Relative frequencies add up to 1.

If relative frequencies are used instead of frequencies:

$$\text{relative frequency density} = \frac{\text{relative frequency}}{\text{class width}}$$

This scales the histogram so that its total area is 1.

Cumulative frequency, quantiles and quartiles

Frequency and cumulative frequency

Given a large set of observations, the first thing to do is to summarise them in a frequency or grouped-frequency table.

For example, the following are the times taken by 50 people to run 100m:

13.5	15.6	16.3	12.3	13.1	14.2	12.4	11.3	14.0	14.6
13.6	14.8	12.7	10.9	11.0	15.0	11.1	15.5	11.3	12.1
12.7	14.6	13.5	15.1	12.1	12.0	14.2	11.4	15.0	13.3
13.2	17.1	16.9	14.2	15.0	13.6	14.8	11.4	14.8	15.7
13.3	13.5	12.9	13.8	13.7	16.2	11.6	13.8	14.2	10.7

GURU TIP
As a rule of thumb, aim for between 5 and 15 groups, with 10 being ideal. Use sensible class widths to decide how many.

A grouped frequency table is used when there are too many different outcomes.

You may have to answer questions like: 'How many people ran the race in less than 15.0 seconds?' For such questions you need to include a column for cumulative frequency in your frequency table.

The above data are thus summarised like this:

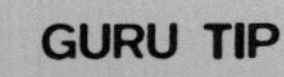

GURU TIP
Cumulative frequency tells you the total number of observations accumulated up to a given point.

Class interval	Frequency	Cumulative frequency
$10.0 \le x < 11.0$	2	2
$11.0 \le x < 12.0$	7	9
$12.0 \le x < 13.0$	8	17
$13.0 \le x < 14.0$	12	29
$14.0 \le x < 15.0$	10	39
$15.0 \le x < 16.0$	7	46
$16.0 \le x < 17.0$	3	49
$17.0 \le x < 18.0$	1	50

For example, this shows that 9 people recorded times of less than 12.0 seconds.

This is what the cumulative frequency graph looks like:

Of course, as the cumulative frequency gives a count up to the upper bound of the class interval, you plot at the upper class boundary. So, for example, you plot the point (15.0,39). Cumulative frequency goes on the y axis.

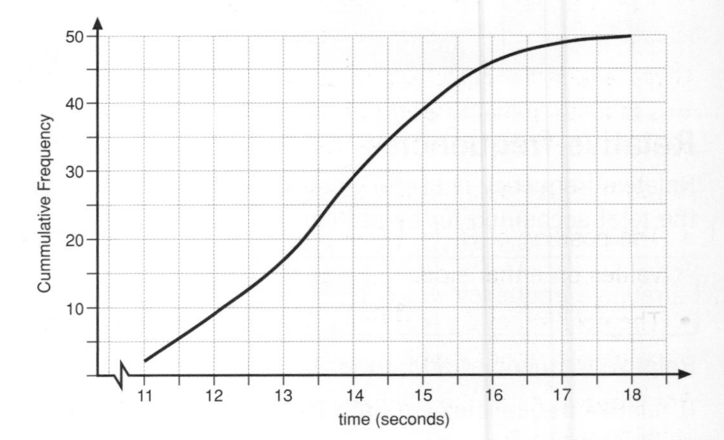

Notice the S shape of the graph.

The graph can now be used to estimate the median. This is the value of x between the 25th and 26th best times. So you draw a line from 25.5 on the y axis across to the curve; then draw a line down to find the estimate of the median.

If you were asked to estimate how many people took more than 15.5 seconds to run the race, you would just draw a line up from 15.5 on the x axis to meet the graph. Read across to find the number of people who took *less than* this time (about 43). Finally take this number away from 50 to find the number of people who took longer than 15.5 seconds (about 7).

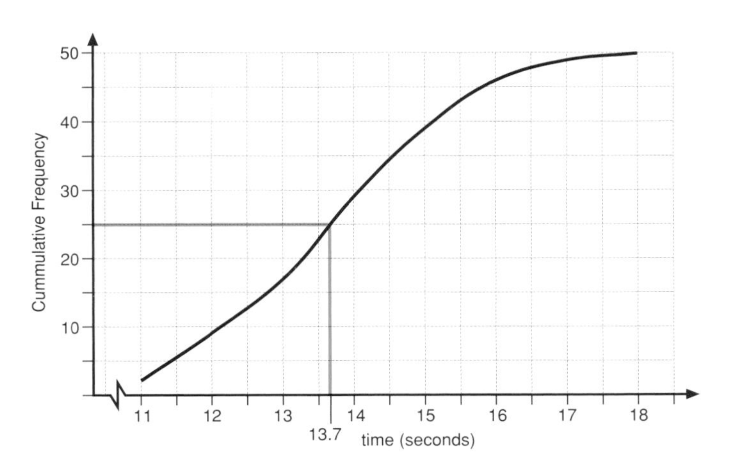

GURU TIP
Draw these lines on your graph.

Looking back at the original data you can see that the actual value is 6. However, you often do not have the original data.

Quartiles, quantiles and interquartile range

You may remember quartiles from your GCSE course. The median finds the middle of the data – that is, 50% of the data will be above the median and 50% below the median. The quartiles divide these halves in half again. So 25% of the data lie below the lower quartile and 25% of the data lie above the upper quartile This leaves 50% of the data between the quartiles.

Q_1 = lower quartile

Q_2 = median

Q_3 = upper quartile

The interquartile range is the difference between the upper and lower quartiles:

IQR = upper quartile – lower quartile = $Q_3 - Q_1$

Given a long list of data, how do you identify the upper and lower quartiles? Well, this is just a generalisation of the method for finding the median. Rather than dividing the data set into two halves, you want to divide it into four quarters.

If you have n items of data arranged in order then:

- The position of Q_1 is $\frac{1}{4}(n + 1)$ – if this is not an integer, take the mean of the values on either side.
- The position of Q_3 is $\frac{3}{4}(n + 1)$ – if this is not an integer, take the mean of the values on either side.

For grouped data, the median and quartiles can be estimated using a cumulative frequency graph.

Quartiles are just one example of **quantiles**. Other examples are deciles (divide into 10 parts) and percentiles (divide into 100 parts).

Box plots, skewness and outliers

Box plots

Box plots are a useful way to show the spread of data, and to compare distributions at a glance.

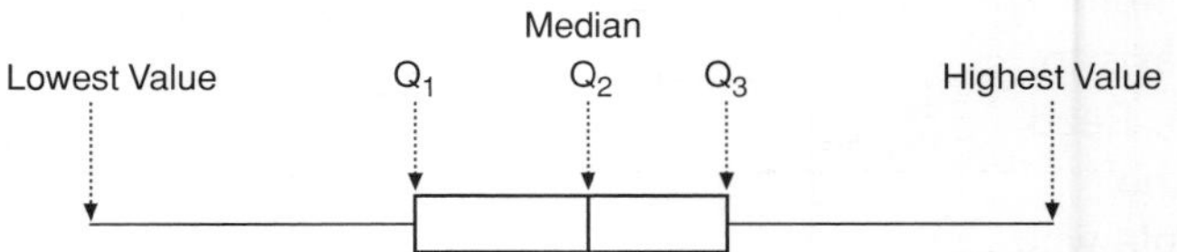

They are sometimes called 'box and whisker' diagrams.

EXAMPLE Draw box plots to compare the following classes' exam results:

Teacher A: 31, 32, 38, 39, 40, 41, 42, 42, 44, 46, 46, 48, 49, 50, 54, 56, 57, 60, 61, 67

Teacher B: 37, 41, 44, 46, 49, 51, 54, 54, 55, 55, 56, 57, 57, 58, 59, 59, 62, 62, 64, 65

SOLUTION First calculate the quartiles. Both sets have 20 numbers.

$\frac{1}{4}(20 + 1) = 5\frac{1}{4}$, so Q_1 = mean of 5th and 6th observations.

For Teacher A, $Q_1 = \frac{1}{2}(40 + 41) = 40.5$

For Teacher B, $Q_1 = \frac{1}{2}(49 + 51) = 50.0$

$\frac{1}{2}(20 + 1) = 10\frac{1}{2}$, so Q_2 = mean of 10th and 11th observations.

For Teacher A, $Q_2 = \frac{1}{2}(46 + 46) = 46.0$

For Teacher B, $Q_2 = \frac{1}{2}(55 + 56) = 55.5$

$\frac{3}{4}(20 + 1) = 15\frac{3}{4}$, so Q_3 = mean of 15th and 16th observations.

For Teacher A, $Q_3 = \frac{1}{2}(54 + 56) = 55.0$

For Teacher B, $Q_3 = \frac{1}{2}(59 + 59) = 59.0$

Now draw the box plots:

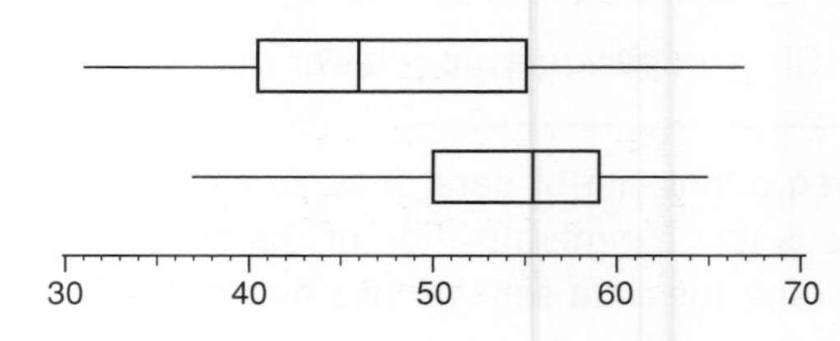

GURU TIP
The quartile may also be read off a cumulative frequency graph. Draw the box plot under the graph using the same scale.

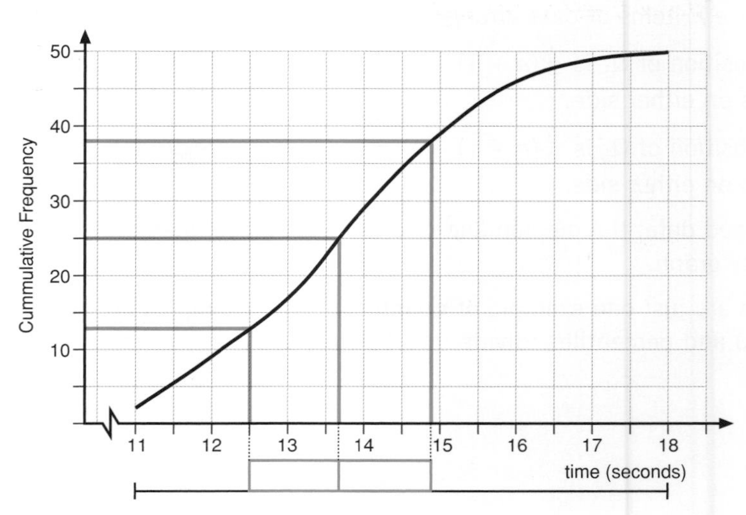

Skewness

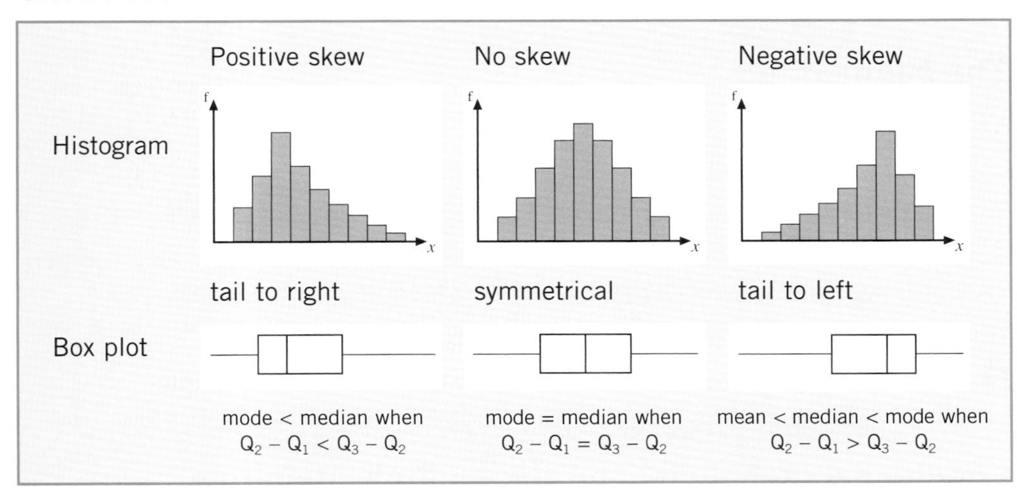

Statistics

Outliers

Outliers or 'extreme points' are unusually small or large observations. It is often said that any observation over one-and-a-half times the IQR away from either Q_3 or Q_1 is an outlier.

When drawing box plots, if there are data values outside this region, they should be marked with an X, but the line should stop at $1\frac{1}{2} \times$ IQR.

EXAMPLE Draw a box plot for the following data set:

11, 12, 21, 22, 22, 23, 24, 25, 26, 27, 28, 28, 28, 29, 30

KEY SKILLS
Use of box plots meets the criteria for Key Skill N3.3.

SOLUTION First find the quartiles.

$Q_1 = 22$

$Q_2 = 25$

$Q_3 = 28$

Find the interquartile range:

$IQR = 28 - 22 = 6$

Identify the outliers:

$Q_1 - 1.5 \times IQR = 22 - 9 = 13$ (outliers 11 and 12)

$Q_3 + 1.5 \times IQR = 28 + 9 = 37$ (no outliers)

Finally draw the box plot:

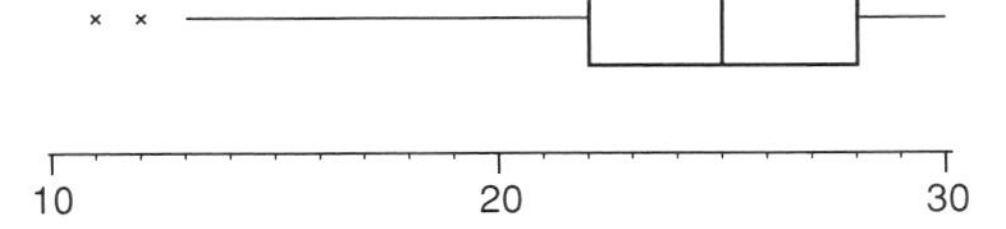

N3.3

Rules of probability

The addition rule

$$P(A \cup B) = P(A) + P(B) - P(A \cap B)$$

WEB TIP
There's more about addition laws on the AS Guru™ website.

Without the last term, anything in both A and B would be counted twice.

So in the example of rolling a die: the probability of getting a number that is either even or prime is the probability getting an even number *plus* the probability of getting a prime number *minus* the probability of getting a number that is both even and prime.

$$P(A \cup B) = \frac{1}{2} + \frac{1}{2} - \frac{1}{6} = \frac{5}{6}$$

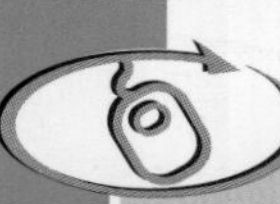

WEB TIP
There's more about Conditional probability on the AS Guru™ website.

Conditional probability and the multiplication rule

Another piece of notation: $P(X \mid Y)$ means 'the probability of X *given that* Y'.

Imagine you are rolling two dice, one green and red. Define the following events:

G = scoring a 6 on the green die

R = scoring a 6 on the red die

E = scoring an even number on the red die

So $P(R \mid E)$ is the probability of scoring a 6 on the red die *given that* the number is even.

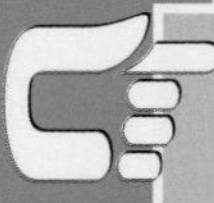

GURU TIP
Note that the order is important: $P(R \mid E) \neq P(E \mid R)$. Think about the example and you'll see why.

The knowledge that the event E has occurred (that is, the number on the red die is even) gives us information about the possible outcomes of the experiment. The sample space has been reduced to those outcomes in E. So the conditional probability is given by the equation:

$$P(R \mid E) = \frac{P(R \cap E)}{P(E)}$$

So the probability of scoring a 6 on the red die *given that* the number is even is $\frac{1}{3}$.

Rearranging the above equation gives: $P(R \cap E) = P(R \mid E) \times P(E)$

WEB TIP
There's more about independence on the AS Guru™ website.

Independent events

Now consider $P(G \mid E)$. Knowing that an even number has been scored on the red die does not tell us anything about the number on the green die. So $P(G \mid E) = P(G)$. The two events are said to be **independent**.

$$\text{Two events A and B are independent} \Leftrightarrow P(A \mid B) = P(A) \qquad (1)$$

The multiplication rule

$$\text{For any two events A and B, } P(A \cap B) = P(A \mid B) \times P(B) \qquad (2)$$

You have used this rule at GCSE when using tree diagrams:

P(B) B P(A | B) A P(A′ | B) A′
P(B′) B′ P(A | B′) A P(A′ | B′) A′

To find the probabilities you multiplied along the branches.

By substituting (1) into (2) we see that:

> For two independent events A and B, $P(A \cap B) = P(A) \times P(B)$

You will be familiar with this equation from GCSE.

Mutually exclusive events

Two events A and B are said to be **mutually exclusive** if they cannot both be the outcome of an experiment. For example, when tossing a coin, the events 'head' and 'tail' are mutually exclusive.

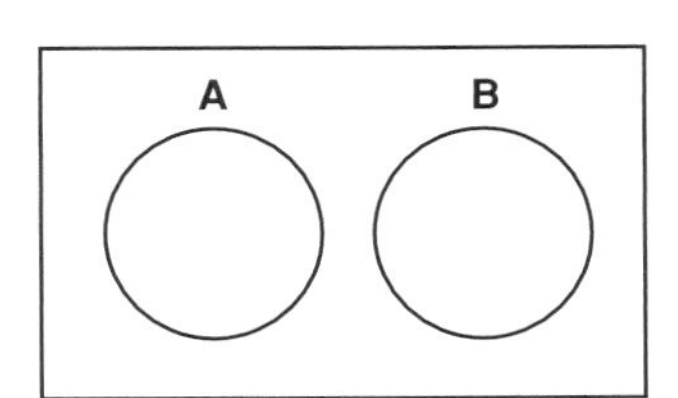

For mutually exclusive events, $P(A \cap B) = 0$

So the addition rule becomes:

> For mutually exclusive events, $P(A \cup B) = P(A) + P(B)$

Statistics

WEB TIP
There's more about two or more events on the AS Guru™ website.

EXAMPLE A medical condition occurs in 5% of the population. A test gives a positive result for 90% of people who have the condition. The test also gives a positive result in 20% of cases where the person does not have the condition.

(a) Find the probability of a randomly selected person getting a positive test result.

(b) Given that Bob got a positive test result, what is the probability that he has the condition?

SOLUTION Let C = 'having the condition', T = 'positive test result'. We know:

$P(C) = 0.05$

$P(T \mid C) = 0.9$

$P(T \mid C') = 0.2$

GURU TIP
Always define exactly what your events are.

(a) A positive test result can arise in two possible ways. Either the person has the condition and tests positive, or the person doesn't have the condition and tests positive.

By the multiplication rule:

$P(C \cap T) = P(T \mid C) \times P(C) = 0.9 \times 0.05 = 0.045$

$P(C' \cap T) = P(T \mid C') \times P(C') = 0.2 \times 0.95 = 0.190$

As these two events are mutually exclusive, you can add the probabilities to find P(T):

The probability of a positive test is $P(T) = P(C \cap T) + P(C' \cap T) = 0.235$.

(b) $P(C \mid T) = \frac{P(C \cap T)}{P(T)} = \frac{0.045}{0.235} = 0.191$

Obviously the test leaves something to be desired!

Counting techniques

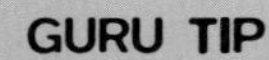

GURU TIP
Check that the probabilities on each set of branches always add up to 1.

WEB TIP
There's more about two or more events on the AS Guru™ website.

Numbers of arrangements

When selecting a number of counters from a bag, or tossing a number of coins, or in any multiple experiments, calculations can be made easier by knowing a little about counting arrangements.

EXAMPLE A bag contains 7 blue counters and 3 red counters. 3 counters are selected at random. What is the probability of getting exactly two red counters?

SOLUTION One method is to draw a tree diagram.

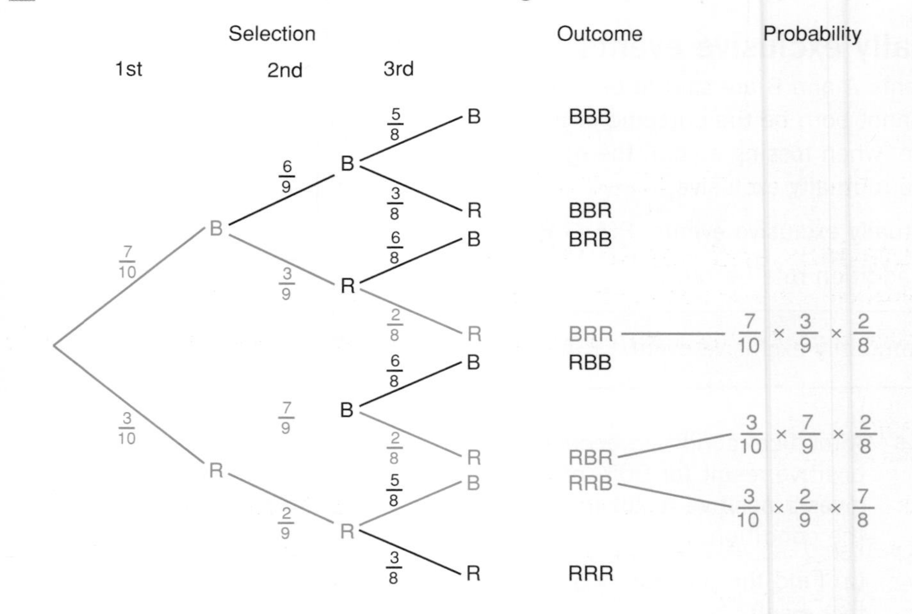

You can now multiply the probabilities as you move through the branches and add up the probabilities at the end of the branches that fit 'exactly two red'.

Actually we do not really need to draw the tree. There are 3 ways of getting exactly two red counters:

1st	2nd	3rd	Probability
R	R	B	$\frac{3}{10} \times \frac{2}{9} \times \frac{7}{8}$
R	B	R	$\frac{3}{10} \times \frac{7}{9} \times \frac{2}{8}$
B	R	R	$\frac{7}{10} \times \frac{3}{9} \times \frac{2}{8}$

Notice that the probabilities are all the same. The denominators are identical, and the numerators are the same but in different orders. So the probability of any one of these outcomes is

$$\frac{3 \times 2 \times 7}{10 \times 9 \times 8} = \frac{7}{120}$$

and the total probability of getting exactly two red counters is

$$3 \times \frac{7}{120} = \frac{7}{40}$$

Tables

You need to be able to use tables to work out probabilities.

EXAMPLE X is the outcome of rolling a red die. Y is the outcome of rolling a blue die. $Z = 2X - Y$. What is $P(Z \leq 4)$?

SOLUTION Draw up a combination table:

		X					
		1	2	3	4	5	6
	1	1	3	5	7	9	11
	2	0	2	4	6	8	10
Y	3	–1	1	3	5	7	9
	4	–2	0	2	4	6	8
	5	–3	–1	1	3	5	7
	6	–4	–2	0	2	4	6

$$P(Z \leq 4) = \frac{\text{(number of outcomes in the event)}}{\text{(total number of outcomes)}} = \frac{21}{36} = \frac{7}{12}$$

Permutations and combinations

A **combination** is a selection in which the order of the items is unimportant.

The number of combinations of r objects chosen from n objects is written as ${}^nC_r$ or $\binom{n}{r}$

$${}^nC_r = \frac{n!}{r!(n-r)!}$$

GURU TIP Check with your teacher whether you need to know these formulae.

EXAMPLE How many ways are there of picking a team of 2 from 6 people?

SOLUTION The order of selection is unimportant. So the number of possible teams is

$${}^6C_2 = \frac{6!}{2!(6-2)!} = \frac{6!}{2!4!} = \frac{6\times5\times4\times3\times2\times1}{2\times1\times4\times3\times2\times1} = 3 \times 5 = 15$$

This could also be solved by common sense. There are 6 ways of choosing the first member and then 5 ways of choosing the second; we have to divide by two as we have counted each pair twice.

A **permutation** is a selection in which the order of the items is important.

The number of permutations of r objects chosen from n objects is written as ${}^nP_r$.

$${}^nP_r = \frac{n!}{(n-r)!}$$

EXAMPLE How many different 2-digit numbers can be formed using 2 of the numbers 1 to 7?

SOLUTION The order of the digits is important. So the number of 2-digit numbers is

$${}^6P_2 = \frac{6!}{(6-2)!} = \frac{6!}{4!} = \frac{6\times5\times4\times3\times2\times1}{4\times3\times2\times1} = 6 \times 5 = 30$$

Random variables 1 – expectation

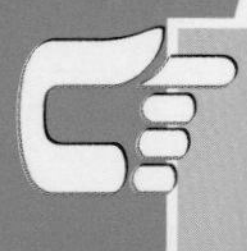

GURU TIP
'X' is a random variable. 'x' is a possible outcome.

The language of random variables

If you roll a die, there are 6 possible outcomes. Call the outcome of the experiment X. The value of X is a number between 1 and 6, but we can't say exactly what it will be. X is a **random variable**.

Given a fair die, the probability of each outcome is $\frac{1}{6}$:

$P(X = 1) = P(X = 2) = \ldots = P(X = 6) = \frac{1}{6}$

'P(X = 1)' is the probability that the die will roll a 1, and so on.

This could be written more succinctly using a **probability function**:

$P(X = x) = \frac{1}{6}\ (x = 1, 2, \ldots, 6)$

A probability function assigns a probability to each of the possible values of a random variable.

Usually, random variables are given capital letters and possible values are given small letters.

This information can be shown as a **probability distribution**. And while we're at it, let's look at the **cumulative distribution function**:

x	1	2	3	4	5	6
$P(X = x)$	$\frac{1}{6}$	$\frac{1}{6}$	$\frac{1}{6}$	$\frac{1}{6}$	$\frac{1}{6}$	$\frac{1}{6}$
$F(x) = P(X \leq x)$	$\frac{1}{6}$	$\frac{2}{6}$	$\frac{3}{6}$	$\frac{4}{6}$	$\frac{5}{6}$	1

As this covers all possible outcomes, the sum of the probabilities must be 1:

$$\sum P(X = x) = 1$$

This means the sum of the probabilities over all possible values of X – read it a symbol at a time.

Success in probability is largely a matter of coping with the notation!

In this example, X is a discrete random variable. That is, it can only take certain values. Continuous random variables can take any value in a given range. (For continuous random variables, it is necessary to integrate a 'probability density function' in order to calculate probabilities – but don't worry, you will not have to do this.)

WEB TIP
There's more about expected value on the AS Guru™ website.

Expectation of a random variable

Remember how to find the mean of a data set from a frequency table? Look at the frequency table below:

x	f	xf
1	4	4
2	2	4
3	1	3
4	1	4
5	1	5
6	1	6
	$\sum f = 10$	$\sum xf = 26$

The mean value of the data is 2.6.

Now imagine that this table shows the outcome of 10 trials of the experiment 'roll a die'.

The mean would give the 'average score'.

Lets look at the data again. If you divide the frequency $f(x)$ for each value x by the total frequency, you get the experimental probability of that outcome. This gives you the estimated probability distribution, $p(x)$.

Multiplying the values by their probabilities and adding gives the mean value.

x	$f(x)$	$\frac{f(x)}{\sum f(x)} = p(x)$	$xp(x)$
1	4	$\frac{4}{10} = 0.4$	0.4
2	2	$\frac{2}{10} = 0.2$	0.4
3	1	$\frac{1}{10} = 0.1$	0.3
4	1	$\frac{1}{10} = 0.1$	0.4
5	1	$\frac{1}{10} = 0.1$	0.5
6	1	$\frac{1}{10} = 0.1$	0.6
	$\sum f(x) = 10$		$\sum xp(x) = 2.6$

GURU TIP
Alwayas use a table when answering questions like this.

As the probabilities of the outcomes are not the same, the die is said to be **biased**.

$p(x)$ is another way of writing $P(X = x)$, that is, the probability that the random variable X takes the value x.

The average value of a random variable X (with probability distribution p) is called the **mean** or **expected value** or **expectation** of X (or p). It is usually written μ or E(X).

$$\mu = E(X) = \sum xp(x)$$

EXAMPLE Find the mean of the following probability distribution.

x	$p(x)$
1	0.2
2	0.7
3	0.1

SOLUTION Add an extra column to the table and calculate $xp(x)$ for each row. Then sum this column to find the mean or expected value $E(X) = \sum xp(x)$.

x	$p(x)$	$xp(x)$
1	0.2	0.2
2	0.7	1.4
3	0.1	0.3
		$\sum xp(x) = 1.9$

So E(X) = 1.9.

Statistics

Random variables 2 – variance

Variance of a random variable

The most important measure of dispersion is the variance.

Remember: $$\text{Var(X)} = \frac{\Sigma (x-\bar{x})^2 f}{\Sigma f} = \frac{\Sigma x^2 f}{\Sigma f} - \bar{x}^2$$

For a probability distribution, $p(x) = \frac{f(x)}{\Sigma f(x)}$, so this becomes

$$\text{Var(X)} = \sum(x-\mu)^2 p(x) = \sum x^2 p(x) - \mu^2$$
where μ is the mean.

GURU TIP
If you are asked for the mean and the variance of a probability distribution, make a table with five columns: x, $p(x)$, $xp(x)$, x^2, $x^2p(x)$. Then calculate $\sum xp(\text{x})$ and $\sum x^2p(x)$.

EXAMPLE Find the variance Var(X) of the following probability distribution:

x	$p(x)$
1	0.2
2	0.7
3	0.1

SOLUTION We calculated in Section 9 that $\mu = \text{E(X)} = 1.9$.

Now add two extra columns, one for x^2 and the other for $x^2p(x)$:

x	$p(x)$	x^2	$x^2p(x)$
1	0.2	1	0.2
2	0.7	4	2.8
3	0.1	9	0.9

So $\sum x^2p(x) = 3.9$

So Var (X) $= 3.9 - 1.9^2 = 0.29$

Expectation and variance of a linear function of a random variable

We sometimes need to work with linear functions of a random variable – for example when coding data.

Suppose 30 students' heights are measured. Define the random variable X as a student's height in metres. The mean or expectation of the heights is E(X), the variance Var(X).

If all the students were now to stand on a table c metres high, their 'heights' would be given by

$\text{Y} = c + \text{X}$

Y is a random variable.

The mean of the new 'heights' is $c + \text{E(X)}$. So $\text{E}(c + \text{X}) = c + \text{E(X)}$.

What has happened to the variance? The variance gives a measure of how spread out the data are. Adding a constant to every value will not affect this.

If every student's height were doubled, the random variable Z = 2X would give the new heights.

Clearly E(Z) = 2E(X). That is, the average height is doubled.

Because the variance is calculated using the *square* of the deviations from the mean, it is increased by a factor of 2^2. So $\text{Var(Z)} = 2^2\text{Var(X)}$.

In general, if a and b are constants and $Y = aX + b$, then

$E(Y) = aE(X) + b$

$Var(Y) = a^2Var(X)$

GURU TIP
As the variance is multiplied by a^2, the standard deviation is multiplied by a.

Statistics

EXAMPLE The random variable X has mean 10 and variance 4. Find the expectation and variance of 3X – 2.

SOLUTION $E(3X - 2) = 3E(X) - 2 = 3 \times 10 - 2 = 28$

$Var(3X - 2) = 3^2Var(X) = 9 \times 4 = 36$

EXAMPLE The random variable X has mean μ and variance σ^2. Find the following in terms of μ and σ^2:

(a) $E(4X)$

(b) $Var(2X - 10)$

(c) $E(X - 4)$

(d) $Var(\frac{1}{2}X - 2)$

SOLUTION

(a) $E(4X) = 4E(X) = 4\mu$

(b) $Var(2X - 10) = 2^2Var(X) = 4\sigma^2$

(c) $E(X - 4) = E(X) - 4 = \mu - 4$

(d) $Var(\frac{1}{2}X - 2) = (\frac{1}{2})^2Var(X) = \frac{1}{4}\sigma^2$

The discrete uniform distribution

A random variable with a finite number of equally likely outcomes is modelled by a **discrete uniform distribution**.

For example, Y = the outcome of rolling a die. Its probability distribution is:

y	$p(y)$
1	$\frac{1}{6}$
2	$\frac{1}{6}$
3	$\frac{1}{6}$
4	$\frac{1}{6}$
5	$\frac{1}{6}$
6	$\frac{1}{6}$

The binomial distribution

Many experiments can be viewed as having two possible outcomes, 'success' and 'failure'. For example, if you call 'heads' when tossing a coin, then this outcome would be a 'success'. If you need to roll a 6 on a die, then this outcome is a 'success' and any other outcome is a 'failure'.

Often we want to know about the outcome of repeated trials in an experiment like this.

EXAMPLE Toss a coin 6 times. Suppose the coin is biased so that $P(H) = \frac{3}{4}$. What is the probability of getting exactly two heads?

SOLUTION Not a question you would want to tackle with a tree diagram!

One favourable outcome is HHTTTT. $P(HHTTTT) = (\frac{3}{4})^2 \times (\frac{1}{4})^4$.

There are many other ways to achieve 2 heads, but they all have the above probability. The total number of orderings of 2 heads and 4 tails is

${}^6C_2 = \frac{6!}{2!4!}$

So $P(\text{2 heads}) = {}^6C_2 \times (\frac{3}{4})^2 \times (\frac{1}{4})^4 = 15 \times \frac{9}{4096} = 0.033$

GURU TIP
Check with your teacher whether you need to cover this topic.

In general, if X is the number of successes in n independent random trials, each with probability p of success, then X has the following probability function:

$P(X = x) = {}^nC_x \, p^x(1-p)^{n-x} \quad (x = 0, 1, \ldots, n)$

This is known as the **binomial distribution**. (It can be derived from the binomial expansion of $(p + (1-p))^n$.)

If a random variable is distributed in this way we write $X \sim \text{Bin}(n,p)$

EXAMPLE A box contains a large number of plain and milk chocolates. There are three times as many plain as milk chocolates. If you select 5 chocolates at random, find the probability of obtaining:

(a) exactly 2 plain chocolates

(b) 2 or more plain chocolates

SOLUTION Let choosing a plain chocolate be a 'success'. So $P(\text{success}) = \frac{3}{4}$. (There are 3 plain to every 1 milk.)

(Note: As the box is large, we can assume that the probabilities are unaffected by any previous selections.)

If X = number of plain chocolates chosen then $X \sim \text{Bin}(5, \frac{3}{4})$.

(a) $P(X = 2) = {}^5C_2 \, (\frac{3}{4})^2 \, (\frac{1}{4})^3 = 10 \times \frac{3^2}{4^5} = 0.088$

(b) $P(X \geq 2) = 1 - P(X = 0) - P(X = 1)$

$= 1 - {}^5C_0 \, (\frac{3}{4})^0 \, (\frac{1}{4})^5 - {}^5C_1 \, (\frac{3}{4})^1 \, (\frac{1}{4})^4$

$= 1 - 1 \times \frac{1}{4^5} - 5 \times \frac{3}{4^5}$

$= 0.984$

GURU TIP
Check that the indices in the denominators always add up to n.

Use of cumulative probability tables

You often need to find probabilities such as $P(X \geq 4)$ or $P(X \leq 6)$. If you have a table of cumulative probabilities, you don't have to calculate lots of individual probabilities and add them up. You can just read the answer from the table.

A biased die is rolled 15 times. If the probability of getting a six is 0.4, use the cumulative frequency table below to find:

(a) the probability of getting fewer than 5 sixes

(b) the probability of getting 8 or more sixes

(c) the probability of getting exactly 6 sixes

r	$P(X \le r)$
0	0.0005
1	0.0052
2	0.0271
3	0.0905
4	0.2173
5	0.4032
6	0.6098
7	0.7869
8	0.9050
9	0.9662
10	0.9907
11	0.9981
12	0.9997
13	1.0000

Let X = the number of sixes. $X \sim \text{Bin}(15,0.4)$.

(a) $P(X < 5) = P(X \le 4) =$ **0.2173**

(b) $P(X \ge 8) = 1 - P(X \le 7) = 1 - 0.7869 =$ **0.2131**

(c) $P(X = 6) = P(X \le 6) - P(X \le 5) = 0.6098 - 0.4032 =$ **0.2066**

We can check (c) by calculating $P(X = 6) = {}^{15}C_6\,(0.4)^6\,(0.6)^9$. You need to be able to write down this formula, but there is nothing to stop you then using tables to calculate the answer.

GURU TIP
Using tables is quicker and less prone to error.

Expectation and variance of a binomial random variable

If $X \sim \text{Bin}(n,p)$ then
$E(X) = np$
$\text{Var}(X) = np(1-p)$

GURU TIP
You will sometimes see this written $\text{Var}(X) = npq$ where $q = 1-p$

Suppose the probability of rain on a given day in Raintown is 0.3, independent of any other day's weather.

(a) How many rainy days do you expect in your two-week holiday in Raintown?

(b) What is the standard deviation of the number of rainy days?

Let X = number of rainy days. $X \sim \text{Bin}(14,0.3)$.

(a) $E(X) = 14 \times 0.3 =$ **4.2 days**

(b) $\text{Var}(X) = 14 \times 0.3 \times 0.7 = 2.94 \Rightarrow \text{s.d.} = \sqrt{2.94} =$ **1.71 days**

The geometric and Poisson distributions

Geometric distribution

GURU TIP
Note that $x = 0$ is not possible. Remember that X is the number of trials including the first success.

Suppose you are playing a board game and you have to roll a six to start. The number of times you need to roll the die until you roll a six is **geometrically distributed**.

If the probability of success is p and X is the number of trials needed to obtain the first success, then X ~ Geo(p).

To have your first success on the xth throw you need x–1 failures followed by one success. As the trials are independent, just multiply the probabilities to find P(X = x).

> If X ~ Geo(p) then
> $P(X = x) = (1-p)^{x-1}p \ (x = 1, 2, \ldots)$
> $E(X) = \frac{1}{p}$
> $Var(X) = \frac{1-p}{p^2}$

The geometric distribution has only one parameter, p, the probability of success. It is usual to write $q = 1-p$ for the probability of a failure.

Here are two useful facts about the geometric distribution:

- probability of no success in the first x trials = the probability of x failures: $P(X > x) = q^x$
- probability of a success in the first x trials = 1 – probability of no success in the first x trials: $P(X \leq x) = 1 - q^x$

EXAMPLE When rolling an unbiased die, what are the probabilities of the following outcomes?

(a) Getting your first six on your 6th throw.

(b) Getting your first six on your 10th throw.

(c) Getting a six in your first 6 throws.

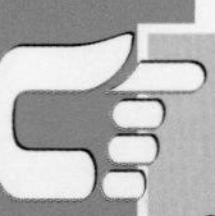

GURU TIP
Don't confuse the situations where the binomial and geometric distributions are appropriate. A binomial distribution has two parameters n and p, and measures the *number of successes*. A geometric distribution has just one parameter p, and measures the *number of trials to the first success.*

SOLUTION Let X = number of throws needed to gain first six. X ~ Geo($\frac{1}{6}$).

$P(X = x) = (\frac{5}{6})^{x-1}(\frac{1}{6}) \ (x = 1, 2, \ldots)$

(a) $P(X = 6) = (\frac{5}{6})^5(\frac{1}{6}) = 0.0670$

(b) $P(X = 10) = (\frac{5}{6})^9(\frac{1}{6}) = 0.0323$

(c) $P(X \leq 6) = 1 - (\frac{5}{6})^6 = 0.6651$

Poisson distribution

The Poisson distribution is used to model situations where:

- Events happen singly and at random in a given interval of time (or space).
- The mean number of occurrences in a given interval is known.

For example, X = the number of phone calls received in an hour. Here the interval is 1 hour. If we know is the mean number of calls per hour then we can model the distribution of X.

If X has a Poisson distribution we write X ~ Po(λ) where λ is mean number of occurrences per unit interval.

If X ~ Po(λ) then

$P(X = x) = \frac{e^{-\lambda}\lambda^x}{x!} (x = 0, 1, 2,...)$

$E(X) = \lambda$

$Var(X) = \lambda$

GURU TIP
Make sure you know how to use cumulative probability tables.

Statistics

Cumulative probability tables are commonly used to answer exam questions on the Poisson distribution.

EXAMPLE Fred is very popular. He receives an average of 4 phone calls per hour. Find the probability that:

(a) Fred receives exactly 6 phone calls in the next hour.

(b) Fred receives more than 6 phone calls in the next hour.

(c) Fred receives no phone calls in the next half-hour.

(d) Fred gets exactly 4 phone calls in each of the next two hours.

SOLUTION Let X = the number of phone calls Fred receives in the next hour. X ~ Po(4).

$P(X = x) = \frac{e^{-4}4^x}{x!} \ (x = 0, 1, 2,...)$

(a) $P(X = 6) = \frac{e^{-4}4^6}{6!} = 0.1042$

(You can also use cumulative probability tables:
$P(X = 6) = P(X \le 6) - P(X \le 5) = 0.8893 - 0.7851 = 0.1042$.)

(b) $P(X>6) = 1 - P(X \le 6) = 1 - 0.8893 = 0.1107$ (using cumulative probability tables).

(c) In a half-hour interval, Fred would expect 2 phone calls. So if Y is the number of calls in the next half-hour then Y ~ Po(2). So

$P(Y = 0) = \frac{e^{-2}2^0}{0!} = e^{-2} = 0.1353$

(d) $P(X = 4) = \frac{e^{-4}4^4}{4!} = 0.1954$. This is the probability of 4 calls in *any* hour. So the probability of 4 calls in the first hour followed by 4 calls in the second hour is $0.1954^2 = 0.0382$.

Distribution of the sum of independent Poisson distributions

If X ~ Po(α), Y ~ Po(β), and X and Y are independent, then X + Y ~ Po(α + β)

KEY SKILLS
The use of multi-stage calculations when using the binomial or Poisson distributions in real-life contexts, satisfies the criteria for Key Skills N3.2.

N3.2

EXAMPLE On a quiet road, the mean number of cars going north is 3 in any 15-minute period and the mean number of cars going south is 1 in any 15-minute period. Find the probability that there will be no cars in the next 15 minutes.

SOLUTION Let X = number of northbound cars in a 15-minute interval and Y = number of southbound cars in a 15-minute interval. So X ~ Po(3) and Y ~ Po(1).

The total number of cars passing in a 15-minute interval is distributed as X + Y ~ Po(4).

$P(X + Y = 0) = \frac{e^{-4}4^0}{0!} = 0.0183$

Correlation

We are often interested in the relationship between two random variables. For example we may be interested in the relationship between the average GCSE score and total A level points of a group of students. For each student we will have a *pair* of observations. The data are **bivariate** and will have a **bivariate distribution**.

Scatter diagrams are a graphical method of investigating the degree of association between two variables. You will have met these, and lines of best fit, at GCSE. There are also numerical ways of measuring the **linear correlation** between two variables.

We would expect there to be some association between GCSE scores and A level points – if one is high we would expect the other to be high. This is called a **positive correlation**.

If one variable, on average, decreases as the other increases, it is called a **negative correlation**. For example, the proportion of people wearing gloves outside would be negatively correlated with the temperature.

If two variables are independent then there is **no correlation**.

The most important numerical summary of a data set is the mean. When asked to draw a scatter diagram you will probably have to calculate the mean for each of the variables and to plot the point $(\overline{x}, \overline{y})$. Drawing a horizontal and a vertical dashed line through this point divides the data into four quadrants.

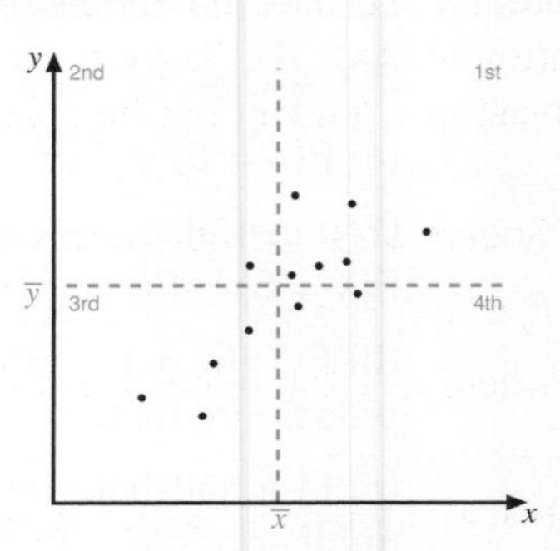

Positive correlation

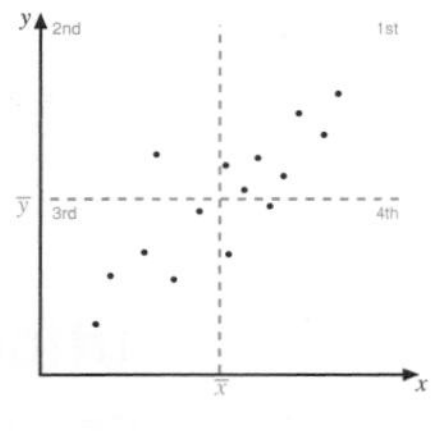

most points in first and third quadrants

Negative correlation

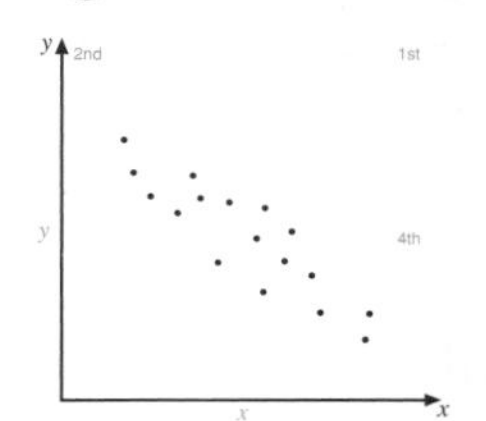

most points in second and fourth quadrants

No correlation

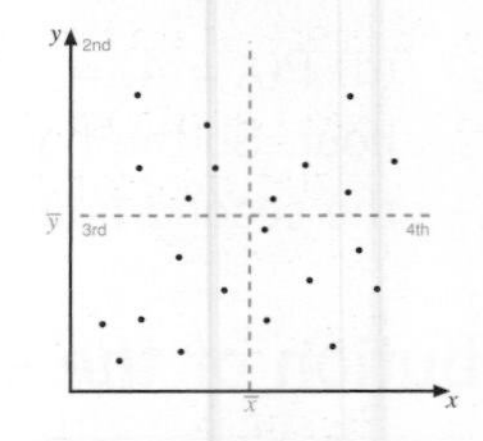

even distribution across all four quadrants

The product-moment correlation coefficient (PMCC)

The PMCC gives a numerical measure of the degree of linear association between two variables. It is defined as

$$r = \frac{\sum(x_i - \overline{x})(y_i - \overline{y})}{\sqrt{\sum(x_i - \overline{x})^2}\sqrt{\sum(y_i - \overline{y})^2}}$$

This is also written

$$r = \frac{S_{xy}}{\sqrt{(S_{xx}S_{yy})}}$$

where

$S_{xy} = \sum(x_i - \overline{x})(y_i - \overline{y})$

$S_{xx} = \sum(x_i - \overline{x})^2$

$S_{yy} = \sum(y_i - \overline{y})^2$

It can be shown that

$S_{xy} = \sum x_i y_i - \frac{\sum x_i \sum y_i}{n}$

$S_{xx} = \sum x_i^2 - \frac{(\sum x_i)^2}{n}$

$S_{yy} = \sum y_i^2 - \frac{(\sum y_i)^2}{n}$

GURU TIP
Different awarding bodies prefer different forms of the formula. Get used to the one yours prefers.

Interpretation

- If $r = 1$, there is a perfect positive linear correlation between the two variables. (In this case all points must lie on a straight line with positve gradient.) The closer r is to 1, the stronger is the positive correlation.
- If $r = -1$, there is a perfect negative linear correlation between the two variables. (In this case all points must lie on a straight line with negative gradient.) The closer r is to -1, the stronger is the negative correlation.
- If $r = 0$, there is no linear correlation between the two variables. The closer r is to 0, the weaker is the linear correlation. However, there may still be some nonlinear relationship.

The modulus of r, $|r|$, gives the degree of correlation – that is, the magnitude of the correlation irrespective of the sign.

Statistics

Spearman's coefficient of rank correlation

This is useful when a variable can be ordered but not measured numerically. For example, two people may be asked to rank a list of films from 1 to 10. (It does not matter whether the 'highest' rank is denoted by 1 or 10 as long as you are consistent. You will not be asked to work with tied ranks.)

GURU TIP
If you are given numerical data, you will need to rank them first.

> Spearman's coefficient of rank correlation is defined as
> $r_s = 1 - \frac{6\sum d^2}{n(n^2-1)}$
> where n is the number of things being ranked and d = rank x – rank y.

- If $r_s = 1$ there is perfect agreement of rankings.
- If $r_s = -1$ the rankings are in exact reverse order.
- If $r_s = 0$ there is no correlation between the rankings.

EXAMPLE Two people were asked to rank 8 flavours of crisp. The results are given in the following table. Calculate Spearman's coefficient of rank correlation.

Flavour	SV	CO	RS	PC	SB	RC	BB	PO
Person X	1	2	4	3	6	5	8	7
Person Y	2	3	4	5	1	6	7	8

SOLUTION First use a table to find $\sum d^2$:

Flavour	SV	CO	RS	PC	SB	RC	BB	PO
d = X – Y	–1	–1	0	–2	5	–1	1	–1
d^2	1	1	0	4	25	1	1	1

$\sum d^2 = 34$

Now use the formula:

$$r_s = 1 - \frac{6\sum d^2}{n(n^2-1)}$$
$$= 1 - \frac{6 \times 34}{8 \times 33}$$
$$= 0.595$$

This shows a positive correlation between the two people's choices.

KEY SKILLS
Calculation and interpretation of a correlation coefficient satisfies the criteria for Key Skills N3.2 and N3.3.

Product-moment correlation coefficient

This is a standard exam question – make sure you can do it.

GURU TIP
Break down the calculation into two steps, and you will have no problems.

EXAMPLE The mean points scored at GCSE, X, and the total A level points, Y, of a sample of 12 students are given below. Calculate the PMCC. Comment on your result.

X	7.5	6.6	6.3	5.1	7.7	5.9	6.1	6.9	7.1	7.2	6.7	6.1
Y	28	21	18	8	26	14	15	22	30	28	24	18

SOLUTION 1. Use a table to find all the terms appearing in the formula for r:

x	y	x^2	y^2	xy
7.5	28	56.25	784	210.0
6.6	21	43.56	441	138.6
6.3	18	39.69	324	113.4
5.1	8	26.01	64	40.8
7.7	26	59.29	676	200.2
5.9	14	34.81	196	82.6
6.1	15	37.21	225	91.5
6.9	22	47.61	484	151.8
7.1	30	50.41	900	213.0
7.2	28	51.84	784	201.6
6.7	24	44.89	576	160.8
6.1	18	37.21	324	109.8
$\sum x = 79.2$	$\sum y = 252$	$\sum x^2 = 528.78$	$\sum y^2 = 5778$	$\sum xy = 1714.1$

2. Calculate r:

$$S_{xy} = \sum xy - \frac{\sum x \sum y}{n} = 1714.1 - \frac{79.2 \times 252}{12} = 50.9$$

$$S_{xx} = \sum x^2 - \frac{(\sum x)^2}{n} = 528.78 - \frac{79.2^2}{12} = 6.06$$

$$S_{yy} = \sum y^2 - \frac{(\sum y)^2}{n} = 5778 - \frac{252^2}{12} = 486$$

$$r = \frac{S_{xy}}{\sqrt{(S_{xx}S_{yy})}} = \frac{50.9}{\sqrt{(6.06 \times 486)}} = 0.938$$

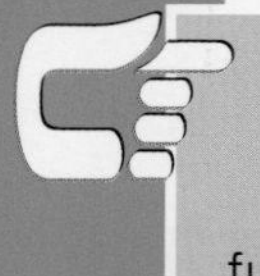

GURU TIP
Many calculators have satistics functions which can help with these calculations.

3. Finally, comment on the significance of your result. The PMCC of 0.938 indicates a strong positive correlation between the students' GCSE and A level results.

You may get a shorter question, giving just the summary statistics, that is, those in the bottom row of the table.

EXAMPLE Find the PMCC for a data set with the following summary statistics:

$\sum x = 336$ $\quad \sum y = 270$ $\quad \sum xy = 17200$

$\sum x^2 = 24989$ $\quad \sum y^2 = 13121$ $\quad n = 8$

SOLUTION

$$S_{xy} = \sum xy - \frac{\sum x \sum y}{n} = 17200 - \frac{336 \times 270}{8} = 5860$$

$$S_{xx} = \sum x^2 - \frac{(\sum x)^2}{n} = 24989 - \frac{336^2}{8} = 10877$$

$$S_{yy} = \sum y^2 - \frac{(\sum y)^2}{n} = 13121 - \frac{270^2}{8} = 4008.5$$

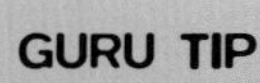

$$r = \frac{S_{xy}}{\sqrt{(S_{xx}S_{yy})}}$$

$$= \frac{5860}{\sqrt{(10877 \times 4008.5)}} = 0.887$$

There is evidence of a strong positive correlation between X and Y.

Coding data when calculating the PMCC

The calculation of the PMCC may be simplified by making the values of the variables smaller. You can add or subtract any number (it often makes sense to subtract the smallest number from all the data), and you can multiply or divide by any number (for example, the highest common factor of the data), without changing the PMCC.

In other words, the PMCC is unaffected by any linear transformation of the data.

N3.1, N3.2, 3.3

EXAMPLE Use the method of coding to find the PMCC for the data x and y below. Comment on your findings.

x	100	90	65	50	45	80	75	70
y	200	220	300	360	340	240	240	320

SOLUTION First transform the data.

For the x data, the smallest number is 45 and the highest common factor is 5. So let $X = \frac{x - 45}{5}$.

For the y data, the smallest number is 200 and the highest common factor is 20. So let $Y = \frac{y - 200}{20}$.

Calculate the PMCC using the transformed data.

X	Y	X^2	Y^2	XY
11	0	121	0	0
9	1	81	1	9
4	5	16	25	20
1	8	1	64	8
0	7	0	49	0
7	2	49	4	14
6	2	36	4	12
5	6	25	36	30
$\sum X = 43$	$\sum Y = 31$	$\sum X^2 = 329$	$\sum Y^2 = 183$	$\sum XY = 93$

Now calculate r:

$$S_{XY} = \sum XY - \frac{\sum X \sum Y}{n}$$

$$= 93 - \frac{43 \times 31}{8} = -73.625$$

$$S_{XX} = \sum X^2 - \frac{(\sum X)^2}{n}$$

$$= 329 - \frac{43^2}{8} = 97.875$$

$$S_{YY} = \sum Y^2 - \frac{(\sum Y)^2}{n}$$

$$= 183 - \frac{31^2}{8} = 62.875$$

$$r = \frac{S_{XY}}{\sqrt{(S_{XX}S_{YY})}}$$

$$= \frac{-73.625}{\sqrt{(97.875 \times 62.875)}} = -0.939$$

There is a very strong negative correlation between x and y.

KEY SKILLS

Selecting this technique to illustrate findings, using and interpreting it, satisfies the criteria for Key Skills N3.1, N3.2 and N3.3.

Regression

If a scatter diagram shows two sets of data to be correlated, we may decide to model the relationship by drawing a straight line through the data.

At GCSE level, choosing the line of best fit required 'judgement'. The regression model allows you to *calculate* the 'best' line through the data.

In science you may have investigated the relationship between the length of a spring and the mass it is supporting. In this experiment, values of the mass are chosen and the corresponding length of the spring is measured. The mass is called the **explanatory** (or **independent**) variable, and the length is called the **response** (or **dependent**) variable.

The regression line would be of length against mass. That is, you would be using your values for the mass to predict or explain the values of the length.

The closeness of the fit depends on how close the points are to the line. The **residuals** are the differences between the actual values of y and the values predicted by the line.

$$r_i = y_i - y$$

The smaller the residuals, the more closely the line fits the data.

> The **least-squares regression line** is the line that minimises the sum of the squares of the residuals.
>
> Its equation is
>
> $y = a + bx$
>
> where $b = \frac{S_{xy}}{S_{xx}}$ and $a = \overline{y} - b\overline{x}$
>
> This is the **regression line of *y* on *x***. It is used for estimating values of y given values of x.

GURU TIP
Different awarding bodies give the formula in different forms. Find out which one yours uses.

If you are given y and asked to estimate x, you will need the regression line of x on y: the formula for this would be obtained by exchanging x and y. In this case the residuals are the differences between the actual and predicted values of x: $r_i = x_i - x$.

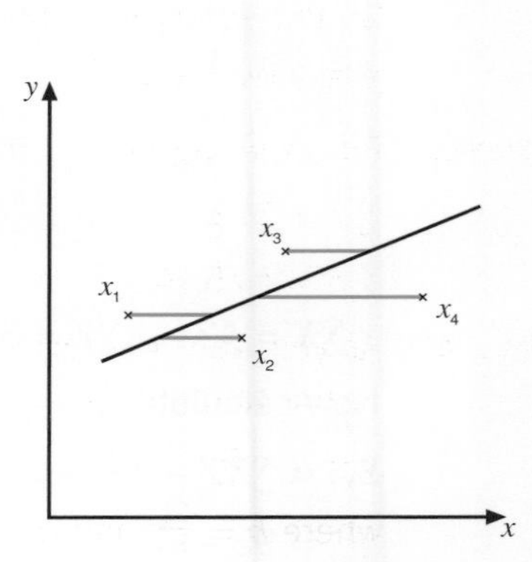

Calculating the equation of the least-squares regression line

EXAMPLE The following table shows the marks obtained by 10 students in their P1 and M1 papers.

Student	A	B	C	D	E	F	G	H	I	J
P1	124	90	107	88	71	60	111	92	103	102
M1	114	92	50	65	65	44	78	81	68	77

Find the equation of the regression line of students' Mechanics marks on their Pure marks.

SOLUTION Let x = the P1 score, y = the M1 score. We require the regression line of y on x. (So no need to interchange variables in the formula.)

Use a table to calculate $\bar{x}$, $\bar{y}$, S_{xy} and S_{xx}.

x	y	x^2	xy
124	114	15376	14136
90	92	8100	8280
107	50	11449	5350
88	65	7744	5720
71	65	5041	4615
60	44	3600	2640
111	78	12321	8658
92	81	8464	7452
103	68	10609	7004
102	77	10404	7854
$\sum x = 948$	$\sum y = 734$	$\sum x^2 = 93108$	$\sum xy = 71709$

$\bar{x} = \frac{948}{10} = 94.8$

$\bar{y} = \frac{734}{10} = 73.4$

$S_{xy} = \sum xy - \frac{\sum x \sum y}{n}$

$= 71709 - \frac{984 \times 734}{10} = 2125.8$

$S_{xx} = \sum x^2 - \frac{(\sum x)^2}{n}$

$= 93108 - \frac{984^2}{10} = 3237.6$

Now

$b = \frac{S_{xy}}{S_{xx}} = 0.66$

$a = \bar{y} - b\bar{y} = 11.2$

So the equation of the regression line is:

$y = 0.66x + 11.2$

GURU TIP
Use your calculator to help you. You can enter the data on a graphics calculator and use linear regression mode to get the equations of regression lines. If you have a graphics calculator, learn how to do this.

EXAMPLE Find the regression line of x on y given:

$\sum x = 948$ $\quad \sum y = 734$ $\quad \sum x^2 = 93108$

$\sum y^2 = 57544$ $\quad \sum xy = 71709$ $\quad n = 10$

SOLUTION As you want the regression line of x on y, you must interchange x and y in the formula. So:

$x = a + by$

where $b = \frac{S_{xy}}{S_{yy}}$ and $a = \bar{x} - b\bar{y}$

$S_{xy} = \sum xy - \frac{\sum x \sum y}{n}$

$= 71709 - \frac{984 \times 734}{10} = 2125.8$

$S_{yy} = \sum y^2 - \frac{(\sum y)^2}{n}$

$= 57544 - \frac{734^2}{10} = 3668.4$

So

$b = \frac{S_{xy}}{S_{yy}} = 0.579$

$a = \bar{x} - b\bar{y} = 94.8 - 0.579 \times 73.4 = 52.3$

So the equation is $x = 52.3 + 0.579y$

KEY SKILLS
Selecting this technique to illustrate findings, using and interpreting it, satisfies the criteria for Key Skills N3.1, N3.2 and N3.3.

GURU TIP
You need to know how to use the formula, otherwise you will not be able to answer questions from summary data.

The normal distribution

The two parameters of the normal distribution

The normal distribution is a continuous probability distribution.

It is of great importance, as many things may be modelled as normal random variables.

If you were to measure the heights of adult men and draw a relative frequency histogram of your results you would get something like this:

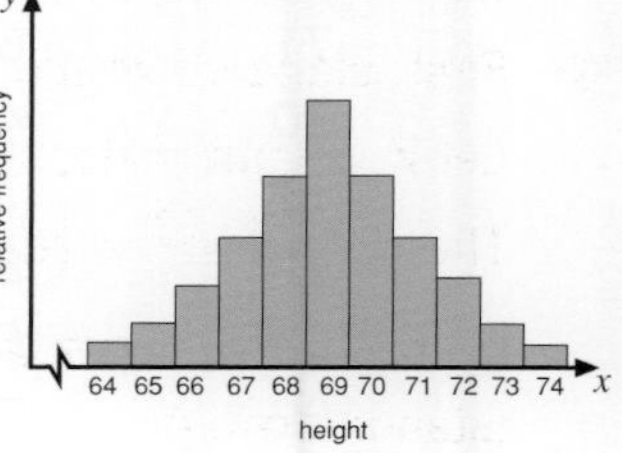

This has approximately the 'bell' shape of the normal distribution:

- It is symmetrical about the mean μ.
- The mode, median and mean are all equal to μ.
- The area under the curve is 1.

μ

The normal distribution is fully described by the two parameters, μ and σ^2. The mean μ determines the position, or location, of the distribution. The variance σ^2 determines the shape, or dispersion, of the distribution. The larger the variance, the wider the bell graph.

If X is a normally distributed random variable with mean μ and variance σ^2, we write $X \sim N(\mu,\sigma^2)$.

The area under the graph up to a certain value of x gives $P(X < x)$:

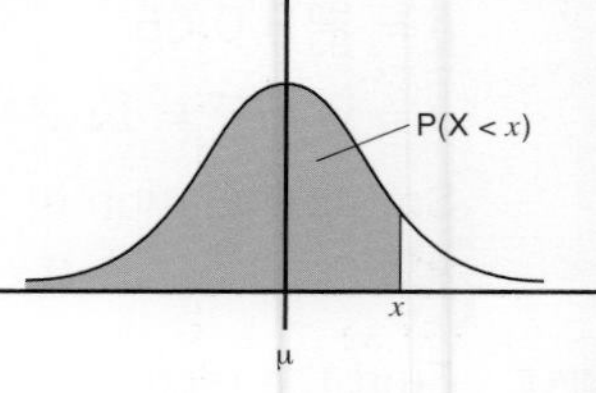

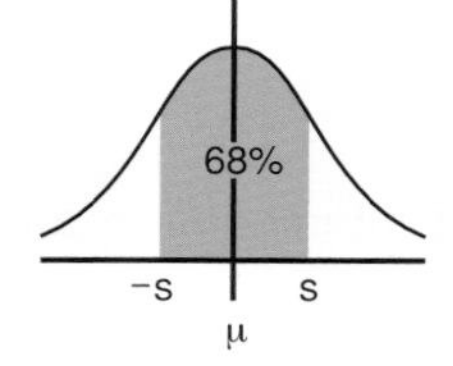

68% of the population lies within 1 s.d. of the mean.

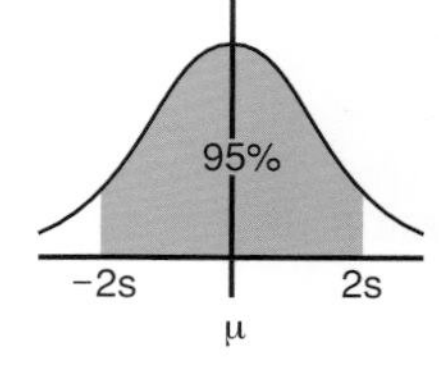

95% of the population lies within 2 s.d. of the mean.

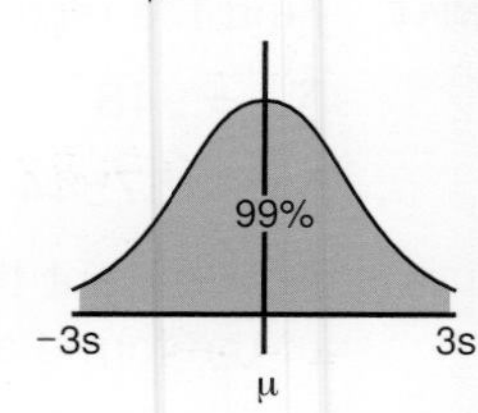

99% of the population lies within 3 s.d. of the mean.

Standardising a normal random variable

To calculate probabilities associated with a normal distribution, you have to transform the distribution to the **standard normal distribution**. This is called **standardising** the random variable.

The standard normal distribution has mean $\mu = 0$ and variance $\sigma^2 = 1$.

To standardise $X \sim N(\mu,\sigma^2)$ you subtract the mean μ and then divide by the standard deviation σ:

$$Z = \frac{X - \mu}{\sigma} \sim N(0,1)$$

Once you have standardised the random variable, you can calculate associated probabilities from tables of $\Phi(z) = P(Z < z)$.

29 A company buys stationery from two suppliers, A and B. It places twice as many orders with A as with B. The probability that supplier A will deliver an order the next day is 0.96. The probability that supplier B will deliver an order the next day is 0.75.

(a) What is the probability that a randomly selected order will not be delivered the next day?

(b) Given that an order has not arrived the next day, what is the probability that it was placed with supplier B?

30 In a card game a player is dealt two cards face down.

(a) Calculate the probability that both cards are aces.

(b) Calculate the probability that the cards are of different suits.

(c) Calculate the conditional probability that both cards are aces given that they are of different suits.

31 A box contains 50 red pens and 25 blue pens. A second box contains 35 red pens and 40 blue ones. A box is chosen at random and a pen is chosen at random from it. Given that the pen is red, what is the probability that it came from the first box?

32 A bag contains 8 identical balls numbered from 1 to 8. A ball is chosen at random from the bag.

A is the event that the ball is numbered 5 or 8.

B is the event that the ball is numbered 2, 3 or 4.

C is the event that the ball is numbered 1, 2, 5 or 6.

Calculate:

(a) $P(A \cap B)$ (b) $P(A \cap C)$ (c) $P(B \cap C)$

Which two events are mutually exclusive?

Which two events are independent?

33 Two events A and B are such that $P(A) = \frac{1}{3}$, $P(B) = \frac{1}{2}$ and $P(B|A) = \frac{1}{4}$.

(a) Calculate $P(B \cap A)$. (b) Draw a Venn diagram to illustrate these data.

(c) Calculate $P(A'|B)$. (d) Calculate $P(A \cup B)$.

Counting techniques

34 A bag contains 5 red pens and 4 blue pens. 3 pens are selected at random from the bag. What is the probability that exactly two of them are red?

35 Two six-sided fair dice are thrown. If both show the same number the score is twice that number, but if they show different numbers then the score is the difference of those numbers. (For example, two 4s scores 8, but a six and a 2 scores 4).

Calculate the probability that the score is (a) 1, (b) 6, (c) 4.

36 A team of 4 people is to be selected at random from a group of 5 men and 4 women. Find the probability that the team contains:

(a) exactly two women (b) at least one man

37 Sixteen students are to be divided into two groups of eight for a practical test. In how many ways can this be done? (The order of the students within the groups does not matter.)

38 (a) How many different five-digit numbers can be formed using the digits 2, 3, 3, 4, 4?

(b) How many of these are even?

(c) How many are divisible by 4?

39 Four cards are selected at random from a standard pack of 52 cards. Calculate the probability that:

(a) all of them are picture cards

(b) two are spades and two are diamonds

Random variables 1 – expectation

40 The probability distribution for X is given below.

x	1	3	4	6
$P(X=x)$	a	0.2	$2a$	a

Calculate:

(a) the value of a (b) E(X) (c) $P(X \leq 4)$

41 Draw the probability distribution table for the number of heads obtained (X) when two fair coins are tossed. Calculate the expected number of heads when two fair coins are tossed.

42 In a game at a village fete, customers throw three dice. If they score a six on one die only they win 20p; if they score a six on two dice they win 40p; and if they score a six on all three dice they win £1. Let X pence be the amount won.

(a) Draw the probability distribution table for X.

(b) Calculate E(X).

(c) If customers pay 20p per turn, what is the expected profit per turn for the stall-holder?

43 A bag contains 3 red discs and 5 black discs, which are identical in size and shape. Three discs are chosen from the bag (without replacement).

(a) What is the expected number of red discs chosen?

(b) Calculate the probability there are at most two red discs in the selection.

44 It is planned to have a stall at the Scout Fair in which participants are asked to pick a card from a standard pack of 52 playing cards.

Find the Lady

Win £1 if you draw the Queen of Diamonds.
Your money back for an Ace or a Spade.

The organisers hope to make at least 25p per customer. What is the minimum they should charge per turn?

Random variables 2 – variance

45 Two tetrahedral dice numbered from 1 to 4 are thrown and the score is the product of the numbers shown on the dice. Calculate:

(a) the expected score (b) the variance of the score

STATISTICS

Data collection and probability

1 4 lower-school pupils, 6 middle-school pupils, 5 sixth-form pupils.

2 Any two of:
- Customers are likely to travel further to pay a first visit to a new supermarket.
- Customers are likely to travel further on a Saturday.
- Not all customers will travel by car.

Survey should be carried out:
- On different days.
- When the supermarket has been open for longer.
- Including motorists and non-motorists.

3 For example, you could allocate the numbers 1 to 300 to the students using registers or an alphabetical list, and use random number tables or a calculator to generate 30 random numbers to choose the students.

It may not be a representative sample: for example, the proportions of girls and boys may not reflect the year group.

4 (a) quantitative, discrete (b) quantitative, continuous

5 (a) $\frac{3}{13}$ (b) $\frac{10}{13}$ (c) $\frac{3}{52}$

6 (a) $\frac{1}{12}$ (b) $\frac{1}{6}$ (c) $\frac{1}{36}$

7 (a) $\frac{1}{13}$ (b) $\frac{1}{4}$ (c) $\frac{1}{52}$ (d) $\frac{4}{13}$

8 (a) $\frac{1}{3}$ (b) $\frac{1}{2}$ (c) $\frac{1}{6}$ (d) $\frac{2}{3}$

9 $\frac{5}{12}$

Numerical methods of summarising data

10 (a) median 70.5, mode 68

(b) mean 70.63, standard deviation 2.83

(c) The mean is representative of the whole data set but it is affected by the one extreme value.

11 mean 6.69, standard deviation 0.71

12 mean 35kg, variance 10.58kg^2

13 mean 5.01cm, standard deviation 0.02cm

Calculating the mean and variance

14 mean 7.25, standard deviation 1.36

15 The upper boundary of the first class is 10 (age is stated in completed years).

mean 38.8 years, standard deviation 23.6 years

16

Class interval	f	x	$u = \frac{x-20.257}{0.002}$	fu	fu^2
20.250–	2	20.251	–3	–6	18
20.252–	3	20.253	–2	–6	12
20.254–	5	20.255	–1	–5	5
20.256–	10	20.257	0	0	0
20.258–	9	20.259	1	9	9
20.260–	6	20.261	2	12	24
20.262–20.264	5	20.263	3	15	45

Answers

$\sum fu = 19$

$\sum fu^2 = 113$

mean 20.25795cm, standard deviation 0.00322cm

17 (a) 7.0, 3.3 (c) 14.0, 6.6

(b) 17.0, 3.3 (d) $7.0a + b$, $3.3a$

Graphical methods of summarising data

18

tens	units
2	5
3	47788
4	1345777
5	02445578
6	179

19 Histogram with class boundaries 119.5, 129.5, 134.5, 139.5, 149.5, 169.5. Bar widths 10, 5, 5, 10, 20 and heights (frequency densities) 0.5, 3.4, 4.0, 1.4, 0.7

20

Class boundaries	15–20	20–25	25–30	30–40	40–50	50–70
relative frequency density (men)	0.0550	0.0375	0.0450	0.0163	0.0100	0.0025
relative frequency density (women)	0.0500	0.0280	0.0400	0.0130	0.0130	0.0075

A higher proportion of younger men play tennis but a higher proportion of older women play tennis.

Cumulative frequency, quantiles and quartiles

21 Upper class boundaries on x axis and cumulative frequencies 2, 8, 21, 37, 64, 72, 76, 80 on y axis.

Median about 95.5kg, interquartile range about 9.5kg. About 26.

22 (b) median 17, IQR 7

(c) mean 17.7, standard deviation 5.7

Box plots, skewness and outliers

23 Firm A: 14, 23, 41. Firm B: 14.5, 27, 34.5.

Firm A: lower boundary 3, upper boundary 59, no outliers.

Firm B: lower boundary 4, upper boundary 41, outlier at 73.

24 Girls: median 47, Q_1 29, Q_3 70, IQR 41.

Boys: median 50, Q_1 36, Q_3 63, IQR 27.

Boys do better on average, but girls have a greater spread of results. More girls scored higher marks.

25 (a) 25, (b) 24.5, (c) 24.2, (d) negative skewness

Rules of probability

26 $\frac{3}{13}$

27 (a) 0.33 (b) 0.61

28 0.46

29 (a) 0.11 (b) 0.76

30 (a) $\frac{1}{221}$ (b) $\frac{13}{17}$ (c) $\frac{1}{169}$

31 $\frac{10}{17}$

32 (a) 0 (b) $\frac{1}{8}$ (c) $\frac{1}{8}$

A and B are mutually exclusive.
A and C are independent.

33 (a) $\frac{1}{12}$ (c) $\frac{5}{6}$ (d) $\frac{3}{4}$

Counting techniques

34 $\frac{10}{21}$

35 (a) $\frac{5}{18}$ (b) $\frac{1}{36}$ (c) $\frac{5}{36}$

36 (a) $\frac{10}{21}$ (b) $\frac{125}{126}$

37 12870 (or 6435)

38 (a) 30 (b) 18 (c) 9

39 (a) 0.00183 (b) 0.0225

Answers

Random variables 1 – expectation

40 (a) 0.2 (b) 3.6 (c) 0.8

41

x	0	1	2
P(X=x)	0.25	0.5	0.25

E(X) = 1

42 (a)

x	0	20	40	100
P(X=x)	$\frac{125}{216}$	$\frac{75}{216}$	$\frac{15}{216}$	$\frac{1}{216}$

(b) 10.19p

(c) 9.81p

43 (a) $\frac{9}{8}$ (b) $\frac{55}{56}$

44 39p

Random variables 2 – variance

45 (a) 6.25 (b) 17.19 (2 d.p.)

46 (a) $\frac{4}{3}$ (b) $\frac{26}{63}$

47

x	1	2	3	4	5	6
P(X=x)	$\frac{1}{8}$	$\frac{1}{8}$	$\frac{1}{8}$	$\frac{1}{8}$	$\frac{1}{8}$	$\frac{3}{8}$

mean 4.13, standard deviation 1.83 (2 d.p.)

48 (a) 10 and 8, (b) 7 and 2

49 (a) 0.1, (b) 3 and 1, (c) 7 and 4

50 E(X) = $4\frac{1}{2}$, Var(X) = $8\frac{1}{4}$

The binomial distribution

51 (a) 0.0536 (b) 0.737 (c) 1

52 (a) 0.0834 (b) 0.7747 (c) 4.8 (d) 2.88

53 (a) 0.107 (b) 0.678 (c) 0.006

54 (a) 9 (b) 0.124

55 (a) 1 and 0.975 (b) 0.0755

56 (a) 9 (b) 1.47

The geometric and Poisson distributions

57 (a) 0.2 (b) 0.102

58 0.226

59 (a) Geometric ($p = \frac{1}{11}$) (b) 0.0621

60 (a) 0.0864 (b) 2.50 and 3.75

61 (a) 0.1680 (b) 0.0258

62 0.594

63 (a) 7 and 7 (b) 0.599 (c) 0.970

64 0.110

Correlation

65 (a) positive (b) none (c) negative

66 0.643, some agreement

67 0.867
Taller students are probably older.

68 0.143

Product-moment correlation coefficient

69 0.833, good positive correlation

70 0.670

71 0.706

72 0.654, some correlation: price provides some guide to performance

Regression

73 (a) $y = 119.5x - 0.842$ (b) 71mm, may not be valid as outside measured range

74 (a) $y = 0.618x + 27.5$ (b) 73

75 (a) $y = 3.67x + 272$ (b) 23.8 million

76 (a) 0.989 (b) $y = 9.28 + 1.34t$ (c) 41.4 million, unreliable as outside range and lower than 1998 figure.

77 (b) $y = 77.8 - 0.639x$ (c) –0.933

The normal distribution

78 (a) 0.76 (b) 0.02 (c) 0.17

79 (a) 0.81 (b) 0.24 (c) 0.96

80 (a) 0.62 (b) 0.31 (c) 0.24 (d) 0.43

81 (a) 7% (b) 31%

82 (a) 11% (b) 6%